Introduction to
CHEMICAL EQUIPMENT DESIGN
Mechanical Aspects

B. C. BHATTACHARYYA

INDIAN INSTITUTE OF TECHNOLOGY
KHARAGPUR

CBS PUBLISHERS & DISTRIBUTORS PVT. LTD.
New Delhi • Bengaluru • Chennai • Kochi • Mumbai • Pune

ISBN: 81-239-0945-4

First Edition: 1985
Reprint: 1991, 1994, 1998, 2000, 2001, 2003, 2005,
2008, 2009, 2011, 2012, 2014, 2015

Copyright © Author

All rights reserved. No part of this book may be reproduced or transmitted in any form or by any means, electronic or mechanical, including photocopying, recording, or any information storage and retrieval system without permission, in writing, from the publisher.

Published by:
Satish Kumar Jain for CBS Publishers & Distributors Pvt. Ltd.,
4819/XI Prahlad Street, 24 Ansari Road, Daryaganj, New Delhi - 110002
delhi@cbspd.com, cbspubs@airtelmail.in • www.cbspd.com
Ph.: 23289259, 23266861, 23266867 • Fax: 011-23243014

Corporate Office: 204 FIE, Industrial Area, Patparganj, Delhi - 110 092
Ph: 49344934 • Fax: 011-49344935
E-mail: publishing@cbspd.com • publicity@cbspd.com

Branches:
- *Bengaluru:* 2975, 17th Cross, K.R. Road, Bansankari 2nd Stage, Benga
  Ph: +91-80-26771678/79 • Fax: +91-80-26771680
  E-mail: cbsbng@gmail.com, bangalore@cbspd.com
- *Chennai:* No. 7, Subbaraya Street, Shenoy Nagar, Chennai - 600030
  Ph: +91-44-26681266, 26680620 • Fax: +91-44-42032115
  E-mail: chennai@cbspd.com
- *Kochi:* 36/14, Kalluvilakam, Lissie Hospital Road, Kochi - 682018
  Ph: +91-484-4059061-65 • Fax: +91-484-4059065
  E-mail: cochin@cbspd.com
- *Mumbai:* 83-C, Dr. E. Moses Road, Worli, Mumbai - 400018
  Ph: +91-9833017933, 022-24902340/41 • E-mail: mumbai@cbspd.cor
- *Pune:* Bhuruk Prestige, Sr. No. 52/12/2+1+3/2,
  Narhe, Haveli (Near Katraj-Dehu Road Bypass), Pune - 411041
  Ph: +91-20-64704058/59, 32342277 • E-mail: pune@cbspd.com

Representatives:
- Hyderabad: 0-9885175004
- Nagpur: 0-9021734563
- Vijayawada: 0-9000660880
- Kolkata: 0-9831437309, 0-905115236
- Patna: 0-9334159340

Printed at:
J.S. Offset Printers, Delhi

PREFACE

THIS book is based upon the lecture notes I prepared during my nine years service on the faculty of the Indian Institute of Technology, Madras. During this period, I had the opportunity to take design classes for undergraduate and postgraduate students of Chemical Engineering and to contribute to an Advanced Summer School on 'Process Equipment Design' organised as a Continuing Education Programme by the Indian Institute of Technology, Madras.

Though stress analysis of chemical equipment is a recent introduction to the Chemical Engineering curriculum, its utility for the better understanding of design is proved beyond doubt. Whatever be the earlier conception, to-day a chemical engineer is expected to be able to make complete design of a piece of chemical equipment. This book is intended to serve this purpose.

The unique feature of this book is that it is written using S.I. (System International) Units only. For those who are not quite familiar with SI units, a chapter on units and conversion factors is included. However, as most of the design equations are derived from the basic stress analysis, any consistent units can be used for numerical solution. There is a wrong conception that code equations are all empirical. It is shown in this book with reference to the Indian Code, how these are developed. This will help the students to understand the applications and limitations of various design equations.

As this book is intended mainly for the senior chemical engineering students, who would be studying elementary theory of mechanics and strength of materials in their earlier years, no separate chapter is included on this. But a chapter defining various terminologies related to stress analysis is given for reference.

The sequence of chapters was selected keeping in view of gradual development of design aspects of a piece of chemical equipment. Selective design examples are added in each chapter to illustrate the application of design equations. Material selection is primary factor in the process of designing chemical equipment. A chapter on this topic gives various aspects of materials of construction. Fabrication techniques and equipment testing are considered to be important to the designer. A chapter is included on these aspects. If economy is neglected, no designer can sustain in his profession. To give an idea how economy could be achieved in sizing process vessels, preliminary approaches of optimization are discussed in the last chapter of this book. Useful design data are included in the Appendices.

I am indebted to my students and colleagues at I.I.T. Madras and I.I.T., Kharagpur for their suggestions. I am especially grateful to Dr. D. Venkateswarlu, Professor-in charge of Chemical Engineering Education Development Centre at I.I.T., Madras for his suggestion to write a text book revising my lecture notes on Process Equipment Design.

I am thankful to Mr. K. D. Chandrasekaran for checking the manuscripts and figures and proof-correction, to Mr. T. Deivasigamani for drawings. Mr. C. Subramanian and Mr. Venkatanarayanan for their secretarial assistance.

Kharagpur

B. C. BHATTACHARYYA

CONTENTS

		PAGE
	Preface	v
1.	Units, symbols and conversion factors	1
2.	Design preliminaries	13
3.	Design of cylindrical and spherical vessels under internal pressure	29
4.	Design of heads and closures	39
5.	Local stresses in process equipment due to discontinuity	57
6.	Compensation for openinings in process equipment	81
7.	Design of non-standard flanges	99
8.	Design of process vessels and pipes under external pressure	127
9.	Design of tall vessels	141
10.	Design of support for process vessels	161
11.	Design of thick walled high pressure vessels	185
12.	Material specifications	197
13.	Equipment fabrication and testing	211
14.	Design of some special parts	223
15.	Optimization	249
	APPENDICES	261
	INDEX	289

CHAPTER 1

UNITS, SYMBOLS AND CONVERSION FACTORS

1.1 INTRODUCTION

Most of the text books till to-day are containing either FPS or MKS units which are not considered to be coherent, rational and comprehensive. A system of units is coherent, if the product or quotient of any two unit quantities in the system is the unit of the resultant quantity ;—for example, in any coherent system, unit area results when unit length is multiplied by unit length ; unit velocity when unit length is divided by unit time ; and unit force when unit mass is multiplied by unit acceleration. The Conference Generale des Poids et Mesures (CGPM) (General Conference on Weights and Measures), in 1954, adopted the rationalized and coherent system of unit based on MKSA units, and in 1960 give it the title 'Systeme' International d' Unites' with the abbreviation 'SI'[1].

The International System of Units (SI) derives all the quantities needed in all technologies from only six basic and arbitrarily defined units. All other units are derived units related to the basic units by definitions. SI system thus possesses features that make it logically superior to any other system and also more convenient as it is coherent, rational and comprehensive. This system is now internationally accepted for expressing the technical units Keeping in mind the future need of the Engineers and students SI units are used throughout this book. Conversion factors are given to facilitate the use of data from other reference books. To maintain uniformity in symbols and signs, IS : 3722-1966 and INSA Booklets on SI Units No. 5 (1970) are followed for this purpose.

To avoid unnecessary increase of the contents the following paragraphs and tables will contain only those units, symbols and conversion factors, which will be used in subsequent chapters of this book.

1.2 BASIC AND DERIVED SI UNITS

For stress analysis of chemical process equipment only 4 basic units are needed. These are listed in Table 1.1.

---

[1] IS : SP : 5-1969 : Guide to the use of SI Units (The Indian Standard Institution, New Delhi).

Table 1.1 Basic SI Units

Physical Quantity	Name of SI Unit	Symbol
Length	metre	m
Mass	kilogram	kg
Time	second	s
Temperature	kelvin	K

The expressions for the derived SI units are stated in terms of the four basic units. In Table 1.2 some of the derived SI units with special names and symbols are listed.

Table 1.2 Derived SI Units with Special Names

Physical Quantity	Name of SI Unit	Symbol
Frequency	hertz	Hz = 1 c/s
Force	newton	N = kg m/s^2
Work, energy	joule	J = N m
Power	watt	W = J/s

All SI units that are not listed in Tables 1.1 and 1.2 may only be expressed in terms of some units from which they are derived. Examples of these are listed in Table 1.3.

Table 1.3 Derived SI Units with Complex Names

Physical Quantity	Name of SI Unit	Symbol
Area	square metre	m^2
Volume	cubic metre	m^3
Density	kilogram per cubic metre	kg/m^3
Velocity	metre per second	m/s
Angular velocity	radian per second	rad/s
Acceleration	metre per second squared	m/s^2

Table 1.3 continued

Table 1.3 continued

Physical Quantity	Name of SI Unit	Symbol
Pressure (stress)	newton per square metre	N/m²
Dynamic viscosity	newton second per square metre	N s/m²
Kinematic viscosity	square metre per second	m²/s
Specific heat	joule per kilogram kelvin	J/kg K
Thermal conductivity	watt per metre kelvin	W/m K

1.3 SELECTION OF DECIMAL MULTIPLES AND SUB-MULTIPLES OF UNITS

Decimal multiples and sub-multiples of SI units are formed by means of the prefixes as shown in Table 1.4.

Table 1.4   Prefixes and their symbols used in SI units

Factor by which Unit is multiplied	Prefix	Symbol
10^6	Mega	M
10^3	kilo	k
10^2	hecto	h
10^1	deca	da
10^{-1}	deci	d
10^{-2}	centi	c
10^{-3}	milli	m
10^{-6}	micro	μ

It is to be noted that the prefixes representing 10 raised to a power which are a multiple of 3 are preferred. To avoid errors in calculations it is recommended that only SI units themselves are used, and not their decimal multiples and sub-multiples. The preferred numerical value range of 0.1—1 000 has been found to be suitable in most applications for expressing the final results.   For example :

Calculated value	Can be better expressed as
12 000 N	12 kN
0.003 94 m	3.94 mm
14 010 N/m²	14.01 kN/m²

This, however, need not be strictly followed.

1.4 SYMBOLS FOR UNITS

Table 1.5 gives the alphabetical list of symbols for physical quantities, with their SI units which are used in subsequent chapters of this book for analyzing the stresses of chemical process equipment.

Table 1.5 List of Symbols

Physical Quantity	Symbol	SI Unit
Acceleration	a	m s^{-2}
Acceleration of free fall	g	m s^{-2}
Angle	$\alpha, \beta, \gamma, \theta, \phi$, etc.	rad, °
Angular frequency	ω	rad s^{-1}
Angular momentum	L	J s
Angular velocity	ω	rad s^{-1}
Area	A, S	m^2
Average speed	$\bar{c}, \bar{u}$	m s^{-1}
Attenuation coefficient	μ	m^{-1}
Breadth	b	m
Bulk modulus	K	N m^{-2}
Coefficient of friction	μ	—
Common temperature	θ, t	°C
Conductivity, thermal	λ, k	W m^{-1} K^{-1}
Cubic expansivity	α, λ	K^{-1}
Density (mass)	ρ	kg m^{-3}
Diameter	d	m
Distance along path	s, L	m
Expansivity, Cubic	α, γ	K^{-1}
Expansivity, Linear	α, λ	K^{-1}
Force	F	N
Frequency	f, ν	Hz, s^{-1}
Frequency, angular	ω	rad s^{-1}
Frequency, rotational	n	s^{-1}
Heat capacity	C	J K^{-1}

Table 1.5 continued

Table 1.5 continued

Physical Quantity	Symbol	SI Unit
Heat flow rate	ϕ	W
Heat, quantity of	θ	J
Height	h	m
Kinetic energy	E_k, T, K	J
Latent heat	L	J
Latent heat, specific	l	$J\ kg^{-1}$
Length	l	m
Linear expansivity	α, λ	K^{-1}
Linear Strain	ε, e	—
Mass	m	kg
Modulus, bulk (modulus of compression)	K	$N\ m^{-2}$
Modulus, shear (modulus of rigidity)	G	$N\ m^{-2}$
Modulus, Young (modulus of elasticity)	E	$N\ m^{-2}$
Moment of couple (torque)	T	$N\ m$
Moment of force (bending moment)	M	$N\ m$
Moment of inertia	I, J	$kg\ m^{-2}$
Momentum	p	$N\ s$
Momentum, angular	L	$J\ s$
Normal stress	σ	$N\ m^{-2}$
Number of turns of coil	N	—
Period	T	s
Poisson ratio	μ, ν	—
Power	P	W
Pressure	p, P	$N\ m^{-2}$
Radius	r	m

Table 1.5 continued

Table 1.5 continued

Physical Quantity	Symbol	SI Unit
Rotational frequency	n	s^{-1}
Shear modulus	G	$N\ m^{-2}$
Specific heat capacity	c	$J\ kg^{-1}\ K^{-1}$
Specific volume	v	$m^3\ kg^{-1}$
Specific weight	γ	$N\ m^{-3}$
Speed	u, v, w	$m\ s^{-1}$
Strain, linear	ε, e	—
Strain, shear	γ	—
Strain, volume	θ	—
Stress, normal	σ	$N\ m^{-2}$
Stress, shear	τ	$N\ m^{-2}$
Temperature, common (celsius)	θ, t	°C, K
Temperature, thermodynamic (absolute)	T, θ	K
Thermal capacity	C	$J\ K^{-1}$
Thermal conductivity	λ, k	$W\ m^{-1}\ K^{-1}$
Thickness	d, δ	m
Time	t	s
Torque	T	$N\ m$
Velocity	u, v, w, c	$m\ s^{-1}$
Viscosity (dynamic)	η, μ	$N\ s\ m^{-2}$
Volume	V, v	m^3
Weight	W, G, P	N
Work	A, W	J
Young's modulus (modulus of elasticity)	E	$N\ m^{-2}$

1.5 RECOMMENDED PRACTICAL UNITS

Although the SI units are preferred, it may not be practical to limit usage to these. For example, it is acceptable to use the metre and millimetre for length measurement (1 m = 10^3 mm), but it would be impractical to limit volume measurement to the cubic metre and cubic millimetre (1 m^3 = 10^9 mm^3). Therefore, the cubic decimetre and cubic centimetre may also be used (1 m^3 = 10^3 dm^3 ; 1 dm^3 = 10^3 cm^3). Table 1.6 gives the list of some recommended practical units which may be used for stating the design problems or expressing the results.

Table 1.6 List of SI Units and Recommended other Units which may be used

Quantity	SI Units	Other Units
Plane angle	rad	degree (°) minute (′) second (″)
Volume	m^3	litre (l) millilitre (ml)
Time	s	day (d) hour (h) minute (min)
velocity	m/s	kilometre per hour (km/h)
Rotational frequency	s^{-1}	revolution per minute revolution per second
Mass	kg	tonne (t) quintal (q)
Density	kg/m^3	t/m^3 ; kg/l g/l ; g/ml
Pressure, normal stress	N/m^2 (pascal)	1 bar = 10^5 N/m^2 1 mbar = 10^2 N/m^2 1 μbar = 10^{-1} N/m^2
Viscosity (dynamic)	N s/m^2	poise (P) = 10^{-1} N s/m^2 centipoise (cP)
Kinematic Viscosity	m^2/s	centistokes (cSt) stoke (St) = 10^{-4} m^2/s
Energy, work	J	kilowatt hour (kW h) 1 kW h = 3.6 MJ

1.6 CONVERSION FACTORS[2]

Conversion factors for normally used quantities into the units of the SI are given in Table 1.7.

Table 1.7 SI Equivalents for existing Units

Quantity	Existing Units	SI Units	Recommended Practical Units	
Acceleration		m/s^2		
	1 cm/s^2	1.000 0 × 10^{-2}	0.01	m/s^2
	1 ft/s^2	3.048 0 × 10^{-1}	0.304 8	m/s^2
	1 m/s^2	1.000 0	1	m/s^2
Area		m^2		
	1 mm^2	1.000 0 × 10^{-6}	1	mm^2
	1 cm^2	1.000 0 × 10^{-4}	100	mm^2
	1 in^2	6.451 6 × 10^{-4}	645.16	mm^2
	1 ft^2	9.290 3 × 10^{-2}	0.092 9	m^2
	1 yd^2	8.361 3 × 10^{-1}	0.836	m^2
	1 m^2	1.000 0	1	m^2
Density		kg/m^3		
	1 kg/l	1.000 0 × 10^3	1	kg/l
	1 lb/ft^3 (gas)	1.601 8 × 10	16.018	kg/m^3
	,, ,, (liquid)	,, ,,	0.016 02	kg/l
	,, ,, (solid)	,, ,,	16.018	kg/m^3
	1 lb/UK gal	9.977 9 × 10	0.099 78	kg/l
	1 lb/US gal	1.198 3 × 10^2	0.119 83	kg/l
	1 g/cm^3	1.000 0 × 10^3	1	kg/l
Energy (Torque)		J		
	1 erg	1.000 0 × 10^{-7}	0.000 1	mJ
	1 ft pdl	4.213 9 × 10^{-2}	0.042 14	J
	1 ft lbf	1.355 8	1.355 8	J
	1 cal	4.186 8	4.186 8	J
	1 kgf m	9.806 7	9.806 7	J
	1 Btu	1.055 1 × 10^3	1.055 1	kJ

Table 1.7 continued

[2] BCE & Process Technology, Vol. 16, No. 9, 1971, pp. 829-832.

UNITS, SYMBOLS AND CONVERSION FACTORS

Table 1.7 continued

Quantity	Existing Units	SI Units	Recommended Practical Units	
		J		
	1 Chu	$1.899\ 1 \times 10^3$	1.899 1	kJ
	1 hp-hr (metric)	$2.647\ 7 \times 10^6$	2.647 7	MJ
	1 hp-hr (British)	$2.684\ 5 \times 10^6$	2.684 5	MJ
	1 kW h	$3.600\ 0 \times 10^6$	3.600 0	MJ
Force		N		
	1 dyne	$1.000\ 0 \times 10^{-5}$	0.01	mN
	1 pdl	$1.382\ 5 \times 10^{-1}$	0.138 25	N
	1 lbf	4.448 2	4.448 2	N
	1 kgf	9.806 7	9.806 7	N
	1 tonnef	$9.806\ 7 \times 10^3$	9.806 7	kN
	1 tonf	$9.964\ 0 \times 10^3$	9.964 0	kN
Frequency		Hz		
	1 c/s	1	1	Hz
Length		m		
	1 μ (micron)	$1.000\ 0 \times 10^{-6}$	1	μm
	1 thou (mil)	$2.540\ 0 \times 10^{-5}$	0.025 4	mm
	1 in	$2.540\ 0 \times 10^{-2}$	25.4	mm
	1 ft	$3.048\ 0 \times 10^{-1}$	304.8	mm
	1 yd	$9.144\ 9 \times 10^{-1}$	0.914 4	m
	1 mile	$1.609\ 3 \times 10^3$	1.609 3	km
Mass		kg		
	1 mg	$1.000\ 0 \times 10^{-6}$	1	mg
	1 grain	$6.480\ 0 \times 10^{-5}$	64.8	mg
	1 lb	$4.535\ 9 \times 10^{-1}$	0.453 6	kg
	1 quintal	$1.000\ 0 \times 10^2$	100	kg
	1 ton (short)	$9.071\ 4 \times 10^2$	0.907 14	tonne
	1 ton (long)	$1.016\ 0 \times 10^3$	1.016	tonne
	1 tonne	$1.000\ 0 \times 10^3$	1	tonne

Table 1.7 continued

Table 1.7 continued

Quantity	Existing Units	SI Units	Recommended Practical Units	
Power		J/s		
	1 ft lbf/min	2.2597×10^{-2}	0.022 6	W
	1 ft lbf/s	1.355 8	0.001 356	kW
	1 m kgf/s	9.806 5	0.009 806	kW
	1 hp (metric)	7.3548×10^2	0.735 48	kW
	1 hp (British)	7.4570×10^2	0.745 7	kW
Pressure (Stress)		N/m^2		
	1 dyne/cm^2	1.0000×10^{-1}	0.001	mbar
	1 N/m^2 (or pascal)	1.000 0	0.01	mbar
	1 kgf/m^2	9.806 7	0.098 07	mbar
	1 mm water	9.806 7	0.098 07	mbar
	1 lbf/ft^2	4.7880×10	0.478 8	mbar
	1 cm water (or gf/cm^2)	9.8067×10	0.980 67	mbar
	1 mbar	1.0000×10^2	1	mbar
	1 matm	1.0133×10^2	1.013 3	mbar
	1 torr (or mm Hg)	1.3333×10^2	1.333 3	mbar
	1 in water	2.4909×10^2	2.490 9	mbar
	1 ft water	2.9891×10^3	0.029 89	bar
	1 in Hg	3.3866×10^3	0.033 87	bar
	1 lbf/in^2 (or psi)	6.8948×10^3	0.068 95	bar
	1 m water	9.8067×10^3	0.098 07	bar
	1 at (or kgf/cm^2 or kp/cm^2)	9.8067×10^4	0.980 67	bar
	1 bar	1.0000×10^5	1	bar
	1 atm	1.0133×10^5	1.013 3	bar
	1 N/mm^2	1.0000×10^6	10	bar
	1 tonf/in^2	1.5444×10^7	154.44	bar

Table 1.7 continued

Table 1.7 continued

Quantity	Existing Units	SI Units	Recommended Practical Units	
Specific Heat Capacity		J/kg K		
	1 cal/gm °C	$4.186\ 8 \times 10^3$	4.186 8	kJ/kg °C
	1 Btu/lb °F	,,	,,	kJ/kg °C
	1 Chu/lb °C	,,	,,	kJ/kg °C
Temperature Difference		K		
	1 K	1.000 0	1	°C
	1 °C	1.000 0	1	°C
	1 °F	5/9	5/9	°C
	1 °R	5/9	5/9	°C
Thermal Conductivity		J/s m K		
	1 Btu/h ft² (°F/in)	$1.442\ 3 \times 10^{-1}$	0.144 23	W/m °C
	1 kcal/h m °C	1.163 0	1.163 0	W/m °C
	1 Btu/h ft °F	1.730 8	1.730 8	W/m °C
	1 cal/s cm °C	$4.186\ 8 \times 10^2$	418.68	W/m °C
Time		s		
	1 min	$6.000\ 0 \times 10$	0.016 67	h
	1 h	$3.600\ 0 \times 10^3$	1	h
	1 day	$8.640\ 0 \times 10^4$	1	day
Velocity		m/s		
	1 ft/h	$8.466\ 7 \times 10^{-5}$	0.084 67	mm/s
	1 ft/min	$5.080\ 0 \times 10^{-3}$	0.005 08	m/s
	1 ft/s	$3.048\ 0 \times 10^{-1}$	0.304 8	m/s
	1 mile/h	$4.470\ 4 \times 10^{-1}$	0.447 04	m/s
Viscosity (dynamic)		N s/m²		
	1 mN s/m² (or cP)	$1.000\ 0 \times 10^{-3}$	1	mN s/m² (cP)
	1 lb/ft h	$4.133\ 8 \times 10^{-4}$	0.413 38	mN s/m²
	1 g/cm s (or poise P.)	$1.000\ 0 \times 10^{-1}$	0.1	N s/m²
	1 lb/ft s	1.488 2	1.488 2	N s/m²
Volume		m³		
	1 in³	$1.638\ 7 \times 10^{-5}$	0.016 39	l
	1 US gal	$3.785\ 3 \times 10^{-3}$	0.003 78	m³
	1 UK gal	$4.546\ 0 \times 10^{-3}$	0.004 55	m³
	1 ft³	$2.831\ 7 \times 10^{-2}$	0.028 32	m³
	1 barrel (petroleum US)	$1.589\ 8 \times 10^{-1}$	0.158 98	m³
	1 lube oil barrel	$2.081\ 9 \times 10^{-1}$	0.208 19	m³
	1 yd³	$7.645\ 5 \times 10^{-1}$	0.764 55	m³

CHAPTER 2
DESIGN PRELIMINARIES

To appreciate the analysis made and decision taken in designing equipment and their parts, it is found necessary to explain the significance of certain terminologies which are used in the subsequent chapters.

2.1 DESIGN CODES

To design pressure vessels and their components, the importance of design codes is well established. Code gives the guidelines for safety design of process equipment which are mostly pressure vessels. It does not directly concern with the economic design of the equipment. Keeping in view "the safety first" principle, it is the designer's responsibility to make his design the most economical.

Besides outlining the standard procedure for safe design, code suggests the preferred dimensions of process vessels and structural components. This helps in large scale production and thus reducing the manufacturing cost.

Most of the industrially advanced countries have a code of their own. A few of the foreign standards widely used in India are:

BS 1500 and 1515
ASME Section VIII
DIN
AD Merkblaetter

India has also produced its first code for unfired pressure vessels—IS : 2825—1969. Though the basis for the development of design equations in all the codes is practically the same, it is interesting to note that the vessel wall thickness which were required in 1960 in various countries for essentially the same service and material, are strikingly different.[1] The wide variation in the value of wall thickness demonstrates the lack of sound technical basis for choosing the design stress.

With the development of the newer materials, facilities for rigorous inspection and testing, greater understanding of the stress distribution in vessels, it is necessary to modify the codes accordingly.

---

[1] Proc. of Adv. Summer School on "Analysis and Design of Pressure Vessels and Piping", M.N.R. Engg. College, Allahabad (India), May, 1972. Fig. 1, p. 6,

2.2 MAXIMUM WORKING PRESSURE

This is the maximum gauge pressure expected under any operating conditions of the process. This pressure limit should not be exceeded for the vessel in operation to avoid the failure of the vessel. Design consideration does not take into account limitless pressure fluctuation inside or outside the vessel. If the statement of the design assignment gives only operating pressure, it will be wise thing for the designer to get the point clarified, whether the pressure mentioned corresponds to the maximum working pressure.

2.3 DESIGN PRESSURE

The pressure used in the design calculations for the purpose of determining the minimum thickness of the various component parts of the vessel is the design pressure, which is to be determined from the following considerations.

2.3.1 For the vessel under internal pressure, the design pressure is obtained by adding a minimum of 5% to the maximum working gauge pressure.[2] This precaution as suggested in the code is necessary considering the time lag to bring down the pressure if it exceeds the maximum value or due to sudden failure of the pressure control device if pressure shoots up, it may take some time to release the pressure by other means. It may also be necessary to set the pressure relief valve at a value higher than the maximum operating pressure to ensure the smooth operation at that condition.

If the customer so desires, the design pressure can be decided to a value more than 1.05 times the maximum operating pressure. This is advisable when there is uncertainly in fixing the maximum working pressure. It is to be noted that the cost of the vessel increases with the increase of design pressure for the same capacity. Except for safety reason, therefore, design pressure should not be taken more than the value specified in the code.

2.3.2 If the static pressure inside the vessel due to liquid head exceeds 5% of the maximum working pressure, design pressure is to be chosen as cited in the following example.

Let, maximum working pressure = 2 bar (gauge) ; liquid height inside the column = 5 m ; then, design pressure to be used at the bottom of the column (say, for water inside) = 2 + (0.5) = 2.5 bar (gauge).[3] If in this problem static pressure would be less than 0.1 bar, the design pressure = 2×1.05 bar = 2.1 bar (gauge) should be used.

---

[2] Indian Standard Code for Unfired Pressure Vessels, IS : 2825–1969, Indian Standards Institution. New Delhi.

[3] AD-Merkblaetter (German).

2.3.3 In the case of vessels subject to inside vacuum or external pressure or both, design pressure can be decided from the following considerations.

Case 1 : On the outside of the vessel atmospheric pressure and inside vacuum corresponds to P bar (absolute) pressure. In this case, design pressure = $(1 - P)$ bar or more and can be upto 1 bar depending upon the reliability of the control system and safety.

Case 2 : Inside atmospheric pressure and external pressure above atmospheric. In such case, design pressure = maximum external gauge pressure + 5% extra as in case of 2.3.1.

Case 3 : Internal pressure below atmosphere and external pressure above atmosphere. Let, P_i = absolute pressure inside the vessel in bar ; and P_o = maximum external gauge pressure in bar ; then, design pressure = $P_o + 0.05\ P_o + (1 - P_i)$ or more and can be upto $(1.05\ P_o + 1)$ bar from safety consideration.

2.3.4 As per IS : 2825-1969 design pressure shall not exceed 200 bar to make use of the code equations.

2.4 DESIGN TEMPERATURE

Determination of appropriate design temperature is important to find the allowable stress value of the material of construction which is temperature dependent. Design temperature is selected from the following considerations :

(a) for unheated parts, — the highest temperature of the stored material :

(b) for the body parts, heated by means of steam, hot water, or similar heating media, — the highest temperature of the heating media, or 10 °C higher than the maximum temperature that any part of the body is likely to attain in course of operation ;

(c) for vessels where direct internal or external heating is employed by means of fire, flue-gas or electricity ; or severe exothermic reaction takes place, — (i) if the vessel is shielded, — the highest temperature of the inside material plus minimum of 20 °C ; (ii) if the vessel is unshielded i.e. for the body parts in direct contact — the highest temperature of the inside material plus minimum of 50 °C. In any case the minimum design temperature shall not be less than 250 °C.

(d) The maximum permissible operating fluid temperatures to be observed for the pressure parts of different materials are given in Table 2.1.

Table 2.1 Fluid Temperature Limitations for Pressure Parts

Material	Max. Permissible Temp. °C
Carbon Steel	540
C-Mo steel	590
Cr-Mo Steel	650
Low Alloy Steel (less than 6% Cr)	590
Alloy Steel (less than 17% Cr)	590
Austenitic Cr-Ni Steel	650
Cast Iron	200
Brass	200

2.5 DESIGN STRESS AND FACTOR OF SAFETY

For design purposes the allowable stress, working stress and design stress are synonymous. If a member is so designed that the maximum stress as calculated for the expected conditions of service is less than some certain value, the member will have a proper margin of security against damage or failure. This certain value is the 'design stress' for the material and condition of service in question. The design stress is always less than the damaging stress which is the least unit stress that will render a member unfit for service before the end of its normal life. This lower magnitude of design stress is suggested because of uncertainty as to the conditions of service, nonuniformity of material, and inaccuracy of stress analysis. The magnitude of this uncertainty is defined as "factor of safety" or design stress factor and this indicates the margin between the design stress and the damaging stress The factor of safety can be reduced in proportion to the certainty with which the conditions of service are known, the intrinsic reliability of the material, the accuracy with which the stress produced by the loading can be calculated, and the degree to which failure is unattended by danger or loss.

As per IS : 2825 — 1969 the design stress values for ferrous and non-ferrous material at the design temperature is determined by dividing the appropriate properties of the material (i e the ultimate stress or the stress at the elastic limit at the design temperature) by factors given in Table 2.2 and taking the lowest value. Some of the design stress values as specified in IS : 2825—1969 are given in Appendix (A).

Table 2.2 Design Stress Factors for Various Materials

Property	Carbon and C-Mn steels	Low Alloy steels	Non-Ferrous Material other than Bolting Material	High Alloy Steels
Certified or specified minimum yield (or 0.2% Proof) stress at design temperature	1.5	1.5	1.5	—
Specified minimum tensile stress at room temperature	3.0	3.0	4.0	2.5
Average stress to produce rupture in 10^5 hours at design temperature	1.5	1.5	1	1.5
Average stress to produce a total creep strain of 1% in 10^5 hours at design temp.	1	1	1	1
Certified 1% proof stress at design temperature	—	—	—	1.5

In the case of castings, the factors given in Table 2.2 is to be divided by a quality factor of 0.75 to take care of unseen defects. This quality factor may be increased to 0.9 provided,

    (a) each casting is radiographically examined at all critical locations and found free from harmful defects;

    (b) all castings are examined at all critical locations using magnetic particle, or penetration fluid procedure;

    (c) castings found to be defective are rejected or repaired to the full satisfaction.

2.6 DESIGN WALL THICKNESS AND MINIMUM ACTUAL WALL THICKNESS

Thickness calculated from the stress consideration alone gives the design wall thickness. For this, standard equations are derived in later chapters. These equations do not concern with the rigidity of the equipment or fabricational feasibility or availability. Design thickness always gives the safe value against induced stresses caused by internal or external pressure or similar other loads. It is not possible to derive any

D—3

simple relationship to predict the required wall thickness for construction of vessels which will be rigid for the given dimensions. Minimum thickness required for rigid construction depends on the size of the vessel, material of construction and type of the reinforcement used (if any). It is not difficult to suggest, however, the minimum wall thickness required from the consideration of fabricational requirement and availability. If the size of the vessel is not large, following magnitudes for the minimum wall thickness may be proposed.

(a) For seamless, welded or brazed pressure vessels the thickness should not be less than 2 mm.

(b) For the pressure vessels made of pure aluminium or soft aluminium alloys, the minimum wall thickness is 3 mm.

(c) For riveted vessels in general minimum wall thickness is 5 mm; but if the container is made with copper, the minimum wall thickness required is 7 mm.

Again, minimum wall thickness requirement varies according to the field of application and for this appropriate standard is to be referred. As an example, IS : 4503-1967 is to be referred for minimum shell thickness for shell and tube Heat Exchanger.

Minimum actual wall thickness of pressure vessel is therefore, a standard available sheet metal or pipe-wall thickness which satisfies the design thickness requirement and takes into account the factors like rigidity, weldability, etc., and also the appropriate allowances for non-uniformity in sheet metal thickness, corrosion-erosion allowances, etc. An experienced design engineer will be able to suggest the appropriate value for the minimum wall thickness, as it involves some uncertainties. Standard dimensions for sheet metals are given in Appendix (B).

2.7 CORROSION ALLOWANCE

Corrosion, erosion or abrasion is a common problem in chemical process equipment. Designer is to give due consideration to this factor while deciding the minimum wall thickness. If the rate of corrosion would be known precisely, it would not be a difficult proposition to provide some extra thickness to safeguard against reduction in thickness to any danger level during the life period of equipment. Unfortunately in most of the practical cases the rate of corrosion cannot be predicted in advance. Only the experienced designer can suggest some reasonable value. However, for the fresh design engineer, following guidelines may be useful.

The additional thickness to be provided for the corrosion allowance may be decided from the following considerations.

(a) For carbon steel and cast iron pressure parts, 1.5 mm on all parts except tubes, in the case of those chemical industries where severe conditions are not expected, and 3 mm in case of petroleum industry or in those chemical industries where severe conditions are expected.

(b) For stainless steel and non-ferrous pressure parts, generally no corrosion allowance is required.

(c) If the wall thickness is more than 30 mm, corrosion allowance may be neglected.

Due consideration shall be given to the selection of materials and to other anti-corrosion measures when a pressure vessel is required to handle an electrolyte which would cause the galvanic corrosion.

2.8 WELD JOINT EFFICIENCY FACTOR

Any joint section in a pressure vessel is considered to be weaker compared to the strength of the rest of the plate metal due to uncertainty related to the quality of joint. A thicker wall thickness is, therefore, required in and around the joint section to improve the strength This extra thickness is not necessary for the rest of the metal body so far as strength is concerned. From practical point of view, however, it will be difficult to procure sheet metal having larger thickness at the edges. This results in overall increase of wall thickness to the value which is recommended for the joint section. This increases the metal consumption which would not be required if the joint quality could be improved by testing the joints by appropriate means and taking necessary corrective measures. It will also be unwise to suggest to undertake the rigorous procedure of testing for all the joints, as this will also incur expenditure. IS code for Unfirmed Pressure Vessels categorizes the type of construction and indicates where full radiography test is mandatory or where spot radiography is sufficient or no radiography is required. Accordingly it specifies the magnitude of weld joint efficiency factor as 1.0, 0.85 or 0.7, 0.6, 0.5 as the case may be. Weld joint efficiency factor 1.0 or 0.85 is commonly used in the design of chemical process equipment.

As per IS : 2825-1969 the welded pressure vessels are divided into three classes, namely, class I, class II and class III. Class I construction is needed, while the vessels are to contain lethal or toxic substances or to be operated under severe conditions For class I Vessels weld joint efficiency factor can be chosen 1.00, as fully radiographed joints are needed for this type of construction. Even a minor defect in the welded joint is not permissible. Structural steels as specified under IS : 226-1962, IS : 961-1962 ; IS : 2062-1962 and IS : 3039-1965 cannot be used for this type of vessels. Except D_o/D_i should not exceed 1.5, there is no limitation on thickness of any parts. Generally this category of vessels are fabricated with double-welded butt-joints fully penetrated. Single-welded butt-joints with backing strip can also be used. But in that case weld-joint efficiency factor should be taken only 0.9 and not 1.00, though the joints are fully radiographed.

Class II vessels are for medium duty operations. Most of the chemical process equipment are falling under this category. Weld joint efficiency factor for this class vessels can be taken 0.85, if the type of joints is double-welded butt-joints with full penetration. In case of single welded butt joints with backing strip, this factor will be 0.8. Materials limitation is same as given for class I. For class II construction the maximum wall thickness including corrosion allowance should not exceed 38 mm. All longitudinal and circumferential joints are required to be spot radiographed.

Class III vessels are for relatively light duties, having plate thickness not in excess of 16 mm excluding corrosion allowance, built for working pressures not exceeding 3.5 bar vapour pressure or 17.5 bar hydrostatic design pressure, at temperatures not exceeding 250 °C and unfired. Class III vessels are not recommended for service at temperatures below 0 °C. No radiography is necessary. All materials including structural steels can be used. Weld joint efficiency factor for Class III vessels should be taken as follows :

(i) Double welded butt joints with full penetration — 0.7

(ii) Single welded butt joints with backing strip — 0.65

(iii) Single welded butt joints with backing strip which may not remain in place — 0.6

(iv) Single welded butt joints without backing strip — 0.55

(v) Single full lap joints for circumferential seams only — 0.5

2.9 DESIGN LOADINGS

In the design of a process vessel, consideration of only design pressure will not produce a safe or workable design. Interaction or combination of various loadings involved in a particular circumstance should be taken into account as some of them induce direct stress and some indirect. All the loadings, however, are not coming to the picture at a time. Relevant loadings are to be considered in a particular situation. Following are the loadings which are experienced by a vessel in some condition or other and the design engineer is to decide the effective ones for a given problem.

The loadings are noted below :

A. Main loadings

(a) Design Pressure including static head :

DESIGN PRELIMINARIES

(b) The weight of vessel and normal contents, or weight of the vessel and maximum contents of water specified for the pressure test;

(c) Wind and earthquake loadings in combination with other loadings.

B. **Supplementary loadings**

(a) Local stresses due to supporting lugs, ring girders, saddles, internal structures or connecting piping.

(b) Shock load due to water hammer or surging of vessel contents;

(c) Bending moment caused by eccentricity of the centre of working pressure relative to the neutral axis of the vessel;

(d) Forces due to the temperature differences, including the effects of differential expansion;

(e) Forces caused by the method of supporting or erection; and

(f) Fluctuating pressure and temperature.

Formal analysis of the effect of the supplementary loadings is only required in cases where it is not possible to demonstrate the adequacy of the proposed design, for example, by comparison with the behaviour of existing similar vessels.

2.10 POISSON'S RATIO

A bar or a plate subjected to axial tension is elongated in the axial direction, but at the same time it undergoes lateral contraction. The ratio of the unit lateral contraction i.e. unit lateral strain to the unit axial elongation or unit axial strain is found to be constant within the elastic limit for a given material. This constant is called Poisson's ratio and is denoted by μ. For structural and pressure vessel steels, μ may be taken as 0.3. The magnitudes of Poisson's ratio for various materials of construction are shown in Appendix (A).

This phenomenon of expansion-contraction is also observed in case of axial compression which causes lateral expansion and the magnitude of μ under the same load condition is same in both the cases, i.e. under axial tension and axial compression.

As per definition of Poisson's ratio and stress-strain relationship, when a rectangular block of some material is subjected to tensile stresses in two perpendicular directions, the elongation in one direction is dependent not only on the stress in this direction but also on the stress in the perpendicular direction. The unit elongation or strain in the direction of the tensile stress σ_1 is σ_1/E. The tensile stress σ_2 in the perpendicular direction will produce lateral contraction in the direction of σ_1 equal to $\mu\sigma_2/E$, so that if both stresses act simultaneously the unit elongation in the direction of σ_1 will be

$$e_1 = \frac{\sigma_1}{E} - \frac{\mu\sigma_2}{E} \qquad \ldots(2.10.1)$$

Similarly, in the direction of σ_2

$$e_2 = \frac{\sigma_2}{E} - \frac{\mu\sigma_1}{E} \qquad \ldots(2.10.2)$$

If one or both of the stresses are compressive, it is necessary only to consider these as negative while determining the corresponding strains from Eqs. 2.10.1 and 2.10.2.

If three tensile stresses, σ_1, σ_2, σ_3 exist on a cube, the strain in the direction of σ_1 is

$$e_1 = \frac{\sigma_2}{E} - \frac{\mu\sigma_2}{E} - \frac{\mu\sigma_3}{E} \qquad \ldots(2.10.3)$$

From Eqs. 2.10.1 and 2.10.2 the stresses existing in a vessel may be determined experimentally from actual strain measurements by means of strain gauge. The expressions for σ_1 and σ_2 as functions of e_1 and e_2 are

$$\sigma_1 = \frac{(e_1 + \mu e_2) E}{1 - \mu^2} \qquad \ldots(2.10.4)$$

$$\sigma_2 = \frac{(e_2 + \mu e_1) E}{1 - \mu^2} \qquad \ldots(2.10.5)$$

2.11 DILATION OF PRESSURE VESSELS

Dilation, or radial growth, of a pressure vessel under internal pressure is an important phenomenon in determining the discontinuity stresses. This is obtained by integrating the hoop strain e_2 in the vessel wall from an axis through the centre of rotation and parallel to a radius.[4] Let δ be the radial growth of a vessel having inner radius r and subjected to internal pressure p. Then,

$$e_2 = \frac{\delta}{r} \qquad \ldots(2.11.1)$$

Substituting for e_2 from Eq. 2.10.2 gives

$$\delta = r\left(\frac{\sigma_2}{E} - \frac{\mu\sigma_1}{E}\right) \qquad \ldots(2.11.2)$$

---

[4] John F. Harvey. "Pressure Vessel Design". East-West Press Pvt. Ltd., New Delhi, p. 34.

DESIGN PRELIMINARIES

If t is the thickness of thin-walled pressure vessel under internal pressure p, it is shown in Ch. 3 that the hoop stress in a cylindrical shell becomes

$$\sigma_2 = \frac{pr}{t}$$

and longitudinal stress

$$\sigma_1 = \frac{pr}{2t}$$

Similarly for spherical vessel these are

$$\sigma_1 = \sigma_2 = \frac{pr}{2t}$$

and for conical shell

$$\sigma_2 = \frac{pr}{t \cos \alpha}$$

$$\sigma_1 = \frac{pr}{2t \cos \alpha}$$

Where α is half the apex angle of the conical shell.

Substituting the values for σ_1 and σ_2 in Eq. 2.11.2 for respective cases, the dilation for vessels of different shapes are found to be as given below.

For cylindrical vessel

$$\delta = \frac{pr^2}{2t E} (2 - \mu) \qquad \ldots(2.11.3)$$

For spherical vessel

$$\delta = \frac{pr^2}{2t E} (1 - \mu) \qquad \ldots(2.11.4)$$

For conical vessel

$$\delta = \frac{pr^2 (2 - \mu)}{2t E \cos \alpha} \qquad \ldots(2.11.5)$$

2.12 MOMENT OF INERTIA

The moment of inertia of an area (also referred as the second moment of an area) with respect to an axis is the sum of the products obtained by multiplying each element of the area dA by the square of its distance from the axis y, and is given by

$$I = \int dA \cdot y^2 \qquad \ldots(2.12.1)$$

The moments of inertia of commonly used areas, with respect to the principal central axes are given in Appendix (C).

2.13 RADIU OF GYRATION

The radius of gyration of an area with respect to a given axis is an important characteristic which is required for the safe design of a column support. If the moment of inertia of an area is expressed in terms of the total area and the square of a distance, then

$$I = \int dA \, y^2 = Ak^2 \qquad \ldots(2.13.1)$$

The quantity k is termed the radius of gyration and is the distance from the axis at which the area could be considered concentrated and still have the same moment of inertia effect. A transposition of this equation gives

$$k = \sqrt{I/A} \qquad \ldots(2.13.2)$$

the form in which it is commonly used. The radius of gyration of commonly used areas are given in Appendix (C).

2.14 SECTION MODULUS

In stress analysis, a quantity equal to the moment of inertia of the area divided by the distance from the principal central axis to the extreme edge of the area is frequently encountered. This quantity is termed the section modulus Z and is given by

$$Z = I/c \qquad \ldots(2.14.1)$$

where c is the distance from the principal central axis to the most remote point of the area. The section modulus largely determines the flexural strength of a column or cylindrical vessel of given material. The values of Z for commonly used areas can be obtained in Appendix (C).

2.15 STRESS CONCENTRATION

Nominally the distribution of elastic stress across the section of a member may be uniform or may vary in some regular manner. When the variation is abrupt, so that within a very short distance the intensity of stress increases greatly, the condition is described as stress concentration. It is usually due to local irregularities in form or discontinuity in shape such as holes for nozzle connections, torus in the formed ends, sharp corners at the junction of cylindrical shell with flat cover and so on. These are termed as "stress raisers", which are not inherent in the number as such, but introduced for some special purpose. The maximum intensity of elastic stress produced by many of the common kinds of stress raisers can be ascertained by mathematical analysis, photoelastic analysis, or direct strain measurements.

Often, the actual stresses caused by concentration are much higher than values computed by the areas involved. The ratio of the true maximum stress to the stress calculated by the conventional formulae, using the net section but ignoring the changed distribution of stress, is the 'factor of stress concentration'. Depending upon the nature of stress raiser, this factor may go as high as 3—5. Occasionally, however, the reduction in strength due to stress concentration is not as serious as might be supposed. The plastic yielding that occurs on overstressing greatly mitigates stress concentration and causes it to have much less influence on breaking strength than might be expected from a consideration of the elastic stresses only. The practical significance of stress concentration therefore depends on circumstances. For ductile metal under static load, it can get adjusted so that the stress from the overstressed area is transferred to the adjacent areas, thereby minimizing the deleterious effect of stress concentration. With brittle materials, on the other hand, stress concentration is always serious. Stress concentrations are especially serious in members subjected to fluctuating loads, and ductile materials under varying stresses often are subject to rapid deterioration when stress concentrations are present.

2.16 THERMAL STRESSES

Whenever the expansion or contraction that would normally result from the heating or cooling of a body is prevented, stresses are developed that are called thermal or temperature stresses There are two different sets of circumstances under which thermal stresses occur. These are :

(a) The form of the body and the temperature conditions are such that there be no stresses except for the constraint from without. In any such case, the stresses may be found by determining the shape and dimensions the body would assume if unconstrained, and then calculating the stresses produced by forcing it back to its original shape and dimensions. For example, if a straight bar is uniformly heated from an initial temperature T_1 to a new temperature T_2 the unit change in dimension is $\alpha (T_2 - T_1)$ where α is the coefficient of thermal expansion and there will be no stresses produced if it is free to expand. If, however, the bar is restricted from expanding at the two ends, but it is free to expand laterally due to Poisson effect, the resulting uniaxial thermal stress becomes :

$$\sigma = - E\alpha (T_2 - T_1) \quad \ldots(2.16.1)$$

If the bar is restricted from two directions i.e., say, there will be no expansion in the x and y directions, the principal strains in both the directions will be equal i.e.,

$$e_1 = e_2 = \alpha (T_2 - T_1)$$

and the principal stresses from Eqs. 2.10.4 and 2.10.5 are :

$$\sigma_1 = \sigma_2 = - \frac{E\alpha (T_2 - T_1)}{1 - \mu} \quad \ldots(2.16.2)$$

When a third restraint is imposed, perpendicular to the *x-y* plane of the bar, the stress becomes :

$$\sigma = -\frac{E\alpha(T_2 - T_1)}{1 - 2\mu} \qquad \ldots(2.16.3)$$

These values of thermal stress for full restraint and hence are the maximum that can be created. The minus sign in the above equations indicates the bar is in compression since its expansion has been restricted. If the bar is prevented from contracting, a tensile stress is produced.

(b) The form of the body and the temperature conditions are such that the stresses are produced even in the absence of any external constraint, simply because of the temperature difference at different parts of the same body. Due to internal expansions or contractions, thermal stresses are induced. A unique example of this type is thick-walled reactor vessel.

2·17 CRITERIA OF FAILURE

To understand the reasons for elastic failure it will be convenient to divide metals into two classes : (1) ductile metals, which exhibit marked plastic deformation at a fairly definite stress like yield point and considerable elongation, and (2) brittle metals, for which the beginning of plastic deformation is not clearly defined and which exhibit practically negligible ultimate elongation. Mild steel is typical of first class, cast iron of the second and an ultimate elongation of 5 per cent has been suggested as the arbitrary dividing line between the two classes of metals.[5]

A ductile metal is usually considered to have failed when it has suffered elastic failure, i.e., when marked plastic deformation has begun. Under simple uniaxial tension this occurs when the stress reaches a value denoted by σ_y, which represents the yield strength, yield point, or elastic limit, according to which one of these is the most satisfactory indication of elastic failure for the material in question. The question arises, when does elastic failure occur under other conditions of stress, such as compression or shear, or a combination of tension, compression and shear ?

The four theories of elastic failure that have received, at various times, the widest acceptance are : (1) the 'maximum stress theory', which states that elastic failure occurs when the maximum tensile stress becomes equal to σ_y, (2) the 'maximum strain theory' which states that elastic failure occurs when the maximum tensile strain, becomes equal to σ_y/E ; (3) the 'maximum shear theory', which states that elastic failure occurs when the maximum shear stress becomes equal to $\frac{1}{2}\tau_y$; (4) the 'theory of constant energy distortion', which states that elastic failure occurs when the principal stresses σ_1, σ_2, σ_3, satisfy the equation[6] :

---

[5] R. Soderberg ; Working Stresses ; J. Appl. Mech., Vol. 2, No. 3, 1935.
[6] R. J. Roark, "Formulas for Stress and Strain", p. 29, 4th edn., McGraw-Hill Book Company.

$$(\sigma_1 - \sigma_2)^2 + (\sigma_2 - \sigma_3)^2 + (\sigma_3 - \sigma_1)^2 = 2\sigma_y^2$$

The criteria discussed above have to do with the elastic failure of material. Such failure may occur locally in a 'member' and do no real damage if the volume of material affected is so small or so located as to have but negligible influence on the form and strength of the member as a whole. Whether or not such local overstressing is significant depends upon the properties of the material and the conditions of service. Fatigue properties, resistance to impact, and mechanical and chemical functioning of the equipment are much more likely to be affected than static strength.

A brittle material cannot be considered to have definitely failed until it has broken, and this takes place either through a tensile fracture when the maximum tensile stress reaches the ultimate strength, or through what appears to be a shear fracture when the maximum compressive stress reaches a certain value.

2.18 ELASTIC STABILITY

Under certain circumstances the maximum load a member will sustain is determined, not by the strength of the material, but by the stiffness of the member. This condition arises when the load produces compression, bending, torsion, or a combination of them which will be proportional to the corresponding deformation. If the body is not elastically stable, the shape of the structure will be altered as a result of insufficient rigidity. It is often the controlling factor when compressive loads are involved. The buckling of a thin cylindrical shell under external pressure, or buckling of a horizontal vessel supported on two saddles, or bending of a thin plate under edge compression, or sagging of a thin circular plate under external loads indicates the lack of elastic stability. Elastic stability for an equipment and its parts is essential for proper functioning of the same.

CHAPTER 3

DESIGN OF CYLINDRICAL AND SPHERICAL VESSELS UNDER INTERNAL PRESSURE

3.1 INTRODUCTION

The shape of most of the chemical process equipment is either cylindrical or spherical or some composite of these. The main function of a process equipment is to contain a media under desired pressure and temperature. In doing so it is also subjected to the action of steady and dynamic support loadings, piping reactions, and thermal shocks which require an overall knowledge of the stresses imposed by these conditions. The final thickness of a process vessel should, therefore, be so chosen that it is not only adequate against the induced stresses caused by internal pressure, but also ensures safety against stresses caused by extraneous agencies as mentioned above.

As it is observed most of the process vessels are having wall thickness small in comparison with their diameter and length, and offer little resistance to bending perpendicular to their surface; but these are very resistant to their plane. The constituents of these surfaces are called "membranes", and the stresses calculated by neglecting bending is called "membrane stresses". In this chapter correlations derived are based on membrane stresses. First, stress distribution across the wall thickness will be assumed to be uniform and then, applying Lame's analysis the formulae given in IS code will be derived.

3.2 THIN WALL VESSELS

The thickness of the wall depends on the operating pressure and temperature. For low pressures, thin walls are used. It is difficult to specify any limiting value for thin wall. An ideal thin wall is that in which membrane stresses are uniform along the thickness. However for practical usage it is suggested[1], if the thickness to diameter ratio is less than 0.1, the shell may be considered as thin wall.

Let a cylindrical vessel of length L, diameter D and thickness t is subjected to an internal pressure p. This will cause hoop stress in the tangential direction and longitudinal

---

[1] Proceedings of Advanced Summer School on Analysis and Design of Pressure Vessel and Piping, organized by M.N.R.E.C., Allahabad (India), May 1972, p. 89.

stress in the axial direction. From Fig. 3.1 the force F, which is balancing radial forces acting on the shell surface, can be found by force balance. Taking the sum of the vertical components of all the forces acting on each half of the shell gives the equilibrium equation:

$$2F = 2 \int_0^{\pi/2} p L \frac{D_i}{2} \sin \theta \, d\theta = pLD_i \qquad \ldots(3.2.1)$$

$$F = \frac{pLD_i}{2} \qquad \ldots(3.2.2)$$

The hoop stress in the shell can be obtained by dividing the force F by the cross-sectional area $A = tL$ of the vessel. Thus,

$$\sigma_\theta = \frac{F}{A} = \frac{pLD_i}{2\,tL} = \frac{pD_i}{2\,t} \qquad \ldots(3.2.3)$$

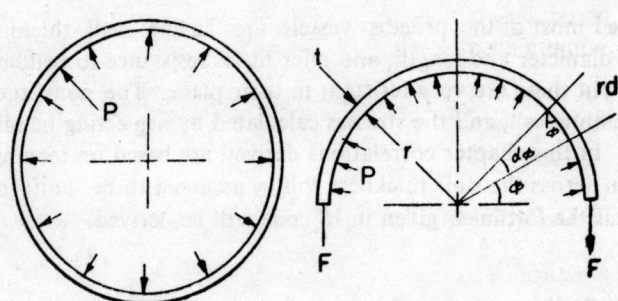

Fig. 3.1 Radial and hoop stresses in thin cylinder

The longitudinal stress can be calculated by equating the total pressure against the end of the cylinder to the longitudinal forces acting on a transverse section of the cylinder, as indicated in Fig. 3.2 giving

$$\pi D t \, \sigma_z = \frac{\pi D_i^2}{4} p \qquad \ldots(3.2.4)$$

$$\sigma_z = \frac{p D_i^2}{4 t D} \qquad \ldots(3.2.5)$$

CYLINDRICAL AND SPHERICAL VESSELS UNDER INTERNAL PRESSURE 31

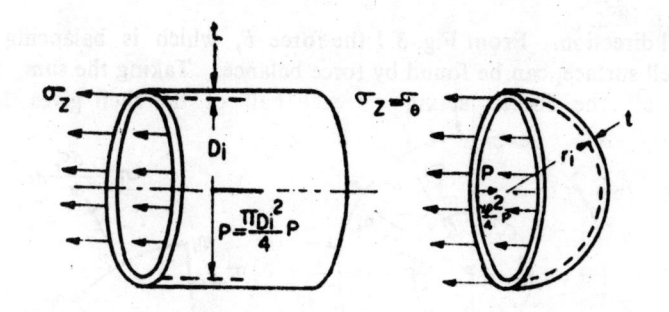

Fig. 3.2 Longitudinal stress in thin cylinder and sphere

In a similar manner, the hoop and longitudinal stresses in a thin sphere subject to internal pressure can be found out to be equal to, and the same as, the longitudinal stress in a cylinder.

$$\sigma_z = \sigma_\theta = \frac{p\, D_i^2}{4t\, D} \qquad \ldots(3.2.6)$$

In all the above equations units are consistent and $D = D_i + t$.

Eq. 3.2.6 is of particular significance in design of process vessels under internal pressure because the minimum absolute stress value $\sigma_z = \sigma_\theta = \sigma_{min}$ is given by a sphere; hence, it is the ideal form stress-wise. Its required thickness for a given set of conditions is one half that necessary for a cylinder.

3.3 DERIVATION OF DESIGN EQUATIONS AS GIVEN IN IS-CODE

In most of the practical cases the wall thickness of process equipment is not so small that the assumption of uniform stress distribution is justified. Actually it is observed when the thickness of the cylindrical vessel is relatively large, the variation in the stress from the inner surface to the outer surface becomes appreciable, and the ordinary membrane or average stress formulae are not a satisfactory indication of the significant stress. In such cases Lame's analysis is used. Lame's solution for thick cylinders is available in the literature.[2] If a cylinder of constant wall thickness is subjected to an internal pressure p_i and external pressure p_o, deformation will be symmetrical about its axis and will not change along its length. In Fig. 3.3 an element mnn_1m_1 of unit length from the wall is considered for force balance. The hoop stress acting on the sides mm_1 and nn_1 is σ_θ. The radial stress normal to the side mn is σ_r, and this stress varies with the radius r in the amount of $(d\sigma_r/dr)\, dr$ over a distance dr. Therefore, the normal radial stress on the side m_1n_1 is

---

[2] John F. Harvey, Pressure Vessel Design, East-West Press Pvt. Ltd., New Delhi, 1969.

$$\sigma_r + \frac{d\sigma_r}{dr} dr \qquad \ldots(3.3.1)$$

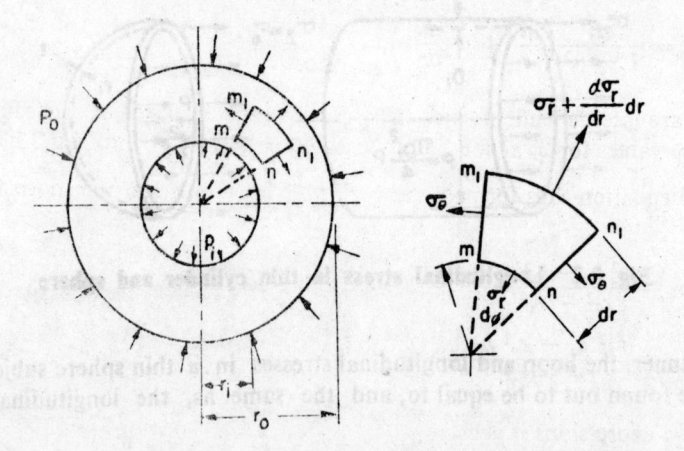

Fig. 3.3 Lame's stress analysis in a thick-walled cylinder

At equilibrium the summation of forces in radial direction yields the following relationship, if the small quantities of high order are neglected,

$$\sigma_\theta - \sigma_r - r \frac{d\sigma_r}{dr} = 0 \qquad \ldots(3.3.2)$$

This equation gives one relation between the stresses σ_θ and σ_r. A second relation can be obtained from the deformation of the cylinder and from the assumption that the longitudinal strain of all fibers is equal. If u is the radial displacement of a cylindrical surface of radius r, the radial displacement of a surface of radius $r + dr$ is

$$u + \frac{du}{dr} dr \qquad \ldots(3.3.3)$$

Therefore, the element mnn_1m_1 undergoes a unit radial elongation or a radial strain of

$$e_r = \left(\frac{du}{dr}\right)\frac{dr}{dr} = \frac{du}{dr} \qquad \ldots(3.3.4)$$

In the circumferential direction the unit elongation or hoop strain of the same element is equal to the unit elongation of the corresponding radius, (see paragraph 2.11), i.e.

$$e_\theta = \frac{u}{r} \qquad \ldots(3.3.5)$$

Then from Eqs. 2.10.4 and 2.10.5 a second set of expressions for the stresses in terms of the strains becomes

CYLINDRICAL AND SPHERICAL VESSELS UNDER INTERNAL PRESSURE

$$\sigma_r = \frac{E}{1 - \mu^2}\left(\frac{du}{dr} + \mu \frac{u}{r}\right) \qquad \ldots(3.3.6)$$

$$\sigma_\theta = \frac{E}{1 - \mu^2}\left(\frac{u}{r} + \mu \frac{du}{dr}\right) \qquad \ldots(3.3.7)$$

These stresses are interdependent since they are expressed in terms of one function u. By substituting the values for σ_r and σ_θ from Eqs. 3.3.6. and 3.3.7 into Eq. 3.3.2 and solving the differential equation thus formed, the general expressions for the normal stresses are given below:

$$\sigma_r = \frac{r_i^2 p_i - r_o^2 p_o}{r_o^2 - r_i^2} - \frac{(p_i - p_o) r_i^2 r_o^2}{r^2 (r_o^2 - r_i^2)} \qquad \ldots(3.3.8)$$

$$\sigma_\theta = \frac{r_i^2 p_i - r_o^2 p_o}{r_o^2 - r_i^2} + \frac{(p_i - p_o) r_i^2 r_o^2}{r^2 (r_o^2 - r_i^2)} \qquad \ldots(3.3.9)$$

From the above expressions it is clear that the maximum value of σ_θ occurs at the inner surface, and maximum σ_r will be the larger of the two pressures, p_i and p_o. These equations are known as the Lame's solution. It is noted that the sum of these two stresses remains constant, hence deformation of all elements in the axial direction is the same as mentioned earlier. The maximum shearing stress at any point in the cylinder is given by

$$\tau = \frac{\sigma_\theta - \sigma_r}{2} = \frac{(p_i - p_o)}{r_o^2 - r_i^2} \times \frac{r_i^2 r_o^2}{r^2} \qquad \ldots(3.3.10)$$

The longitudinal stress is generally small and is given by

$$\sigma_z = \frac{p_i r_i^2 - p_o r_o^2}{r_o^2 - r_i^2} \qquad \ldots(3.3.11)$$

For a cylinder subjected to only internal pressure, p, the expression for the stresses are,

$$\sigma_r = \frac{r_i^2 p}{r_o^2 - r_i^2}\left(1 - \frac{r_o^2}{r^2}\right) \qquad \ldots(3.3.12)$$

$$\sigma_\theta = \frac{r_i^2 p}{r_o^2 - r_i^2}\left(1 + \frac{r_o^2}{r^2}\right) \qquad \ldots(3.3.13)$$

It should be noted that σ_θ is greater than σ_r and it is the maximum at the inner periphery of the cylinder. Therefore, for the design purposes, the hoop stress at $r = r_i$ should be taken as critical or allowable stress, i.e.

$$(\sigma_\theta)_{r = r_i} = f = \frac{p(r_i^2 + r_o^2)}{r_o^2 - r_i^2} \qquad \ldots(3.3.14)$$

where, f stands for allowable stress value of the material of construction. The chemical process equipment is generally fabricated with sheet metals necessitating the end-joints, which reduces the allowable strength of the material with a factor called joint efficiency factor. If J denotes joint efficiency factor, then,

$$fJ = \frac{p(r_o^2 + r_i^2)}{r_o^2 - r_i^2} \qquad \ldots(3.3.15)$$

$$p = fJ\left(\frac{r_o^2 - r_i^2}{r_o^2 + r_i^2}\right) \qquad \ldots(3.3.16)$$

If t is the minimum wall thickness required for the shell,

$$p = fJ \frac{(r_i + t)^2 - r_i^2}{(r_i + t)^2 + r_i^2}$$

$$= \frac{fJt(t + 2r_i)}{(2r_i^2 + 2r_i t + t^2)}$$

$$= \frac{fJt}{[r_i + \frac{t(r_i + t)}{2r_i + t}]}$$

or, $$p = \frac{fJt}{r_i + t\frac{1 + \frac{t}{r_i}}{2 + \frac{t}{r_i}}} \qquad \ldots(3.3.16)$$

Eq. 3.3.16 is the basis for all design equations specified in the codes for internal pressure vessels. The above equation can also be written as,

$$p = \frac{2fJt}{D_i + t\frac{1 + 2t/D_i}{1 + t/D_i}} \qquad \ldots(3.3.17)$$

Limitations stipulated in IS : 2825—1969 for using design equations indicate that

$$D_o/D_i \leqslant 1.5$$

where, D_o and D_i are the outer and inner diameter of the vessel respectively. From the above, maximum value of

$$t/D_i = 0.25$$

It is evident that for the values of t/D_i between 0 to 0.25, the magnitude of the expression $\frac{1 + 2t/D_i}{1 + t/D_i}$ varies from 1.0 to 1.2. As an approximation, IS : 2825—1969 has accepted 1.0

for this expression for t/D_i not exceeding 0.25. The final expressions from the hoop stress consideration are, therefore, given as follows for the cylindrical pressure vessels[3].

$$p = \frac{2fJt}{D_i + t} = \frac{2fJt}{D_o - t} \qquad \ldots (3.3.18)$$

$$t = \frac{pD_i}{2fJ - p} = \frac{pD_o}{2fJ + p} \qquad \ldots (3.3.19)$$

In the case of spherical vessel under internal pressure, the induced hoop and longitudinal stresses are equal and the magnitude is same as the longitudinal stress in a cylinder. The general expression for longitudinal stress in a cylinder is given in Eq. 3.3.11. For internal pressure only this becomes

$$\sigma_z = \frac{p\,r_i^2}{r_o^2 - r_i^2} \qquad \ldots (3.3\,20)$$

For sphere,

$$\sigma_z = \sigma_\theta = fJ = \frac{pr_i^2}{r_o^2 - r_i^2} \qquad \ldots (3.3.21)$$

or $\quad fJ = \dfrac{pD_i^2/4}{t(2r_i + t)}$

or $\quad 4fJt = \dfrac{pD_i^2}{(D_i + t)}$

or $\quad \dfrac{4fJt}{D_i + t} = \dfrac{p}{\left(1 + \dfrac{t}{D_i}\right)^2} \qquad \ldots (3.3.22)$

IS: 2825—1969 has approximated $\left(1 + \dfrac{t}{D_i}\right)$ to unity for t/D_i varies between 0 and 0.25. This gives conservative estimation for higher ratios of t/D_i. The code expressions are therefore,

$$p = \frac{4fJt}{D_i + t} = \frac{4fJt}{D_o - t} \qquad \ldots (3.3.23)$$

$$t = \frac{pD_i}{4fJ - p} = \frac{pD_o}{4fJ + p} \qquad \ldots (3.3.24)$$

It is to be noted carefully that the expressions given in Eqs. 3.3.19 and 3.3.24 for wall thickness t, indicate the theoretically required minimum wall thickness to withstand

---

[3] IS: 2825—1969, Code for unfired pressure vessels, Indian Standards Institution, New Delhi.

the effect of internal pressure p safely. These do not include any extra allowances necessary to take care of corrosion, non-uniformity in sheet metal thickness and reduction of thickness during forming. After adding the appropriate allowances to the theoretically calculated wall thickness, the final value is to be chosen from the standard thickness of the sheet metal available in the market and this should be equal to or higher than the calculated value. For standard dimensions see Appendix (B).

Design Example 3.1: A process vessel is to be designed for the maximum operating pressure of 500 kN/m². The vessel has the nominal diameter of 1.2 m and tangent to tangent length of 2.4 m. The vessel is made of IS : 2002—1962 Grade 2B quality steel having allowable design stress value of 118 MN/m² at working temperature. The corrosion allowance is suggested to be 2 mm for the life span expected for the vessel. The vessel is to be fabricated according to class 2 of Indian Standard specifications which stipulate the weld joint efficiency of 0.85.

(a) What will be the standard plate thickness to fabricate this vessel?

(b) If a spherical vessel having the same diameter and thickness is fabricated with the same quality steel, what maximum internal pressure the sphere will withstand safely?

Solution: (a) To determine the minimum wall thickness required excluding any extra allowances to withstand the induced membrane stress due to operating pressure inside the cylindrical vessel, Eq. 3.3.19 is applicable assuming that t/D_i will not exceed 0.25 which is to be checked later.

$$t = \frac{p\,D_o}{2fJ + p} \qquad \ldots (3.3.19)$$

Here, t = minimum wall thickness without corrosion allowance, m

p = design pressure, N/m² (According to IS : 2825—1969, to decide the design pressure, a minimum of 5% extra is to be added to the maximum working pressure, if a higher value is not asked by the party). In this problem,

$p = 1.05 \times 500 \times 1\,000$ N/m²

$ = 525 \times 1\,000$ N/m²

D_o = outer diameter of the vessel, m (This is the same as nominal diameter according to Indian Standard).

$D_o = 1.2$ m

f = allowable design stress for the material specified, N/m²

$ = 118 \times 10^6$ N/m²

J = weld joint efficiency factor,
  = 0.85
$fJ = 100.3 \times 10^6$ N/m²

Substituting the values in Eq. 3.3.19,

$$t = \frac{525 \times 10^3 \times 1.2}{2 \times 100.3 \times 10^6 + 525 \times 10^3} \text{ m}$$

$$= 3.13 \times 10^{-3} \text{ m}$$

$$= 3.13 \text{ mm}$$

Adding corrosion allowance to this, minimum actual wall thickness required to be

$$t_a' = t + c = 3.13 + 2 = 5.13 \text{ mm}$$

Standard thickness (nearest higher) available is 6 mm. Therefore, to fabricate this vessel, sheet metal of 6 mm thick is to be used.

(b) To determine the maximum internal pressure for a spherical vessel of $D_0 = 1.2$ m and $t = t_a - c = 6 - 2 = 4$ mm, Eq. 3.3.23 is to be used. From this equation, first design pressure is to be evaluated.

$$p = \frac{4fJt}{D_o - t} \qquad \ldots (3.3.23)$$

Substituting the values,

$$p = \frac{4 \times 118 \times 10^6 \times 0.85 \times 4 \times 10^{-3}}{1.2 - 4 \times 10^{-3}} \text{ N/m}^2$$

$$= 1.34 \times 10^6 \text{ N/m}^2$$

Maximum internal pressure is obtained by dividing p by 1.05. Therefore

$$p_i \text{ (max)} = \frac{1.34 \times 10^6}{1.05} \text{ N/m}^2$$

$$= 1.276 \times 10^6 \text{ N/m}^2$$

The spherical vessel having the same diameter and thickness will withstand a pressure of 1.276 MN/m² safely.

In both the cases t/D_i is same, and that is

$$t/D_i = \frac{4}{1\,200 - 2 \times 4}$$

$$= 0.003\,4$$

The ratio is much less than 0.25 and hence the equations are correctly applied.

CHAPTER 4

DESIGN OF HEADS AND CLOSURES

4.1 INTRODUCTION

The ends of a cylindrical process vessel are to be closed before putting into operation. This is done by means of heads and closures, which are of different shapes. The vessels are usually provided with the following types of heads :

      (a) Flat head

      (b) Standard dished or torispherical head

      (c) Ellipsoidal

      (d) Hemispherical

      (e) Conical or toriconical head

Some other forms are also in use.

Various forms in use are shown in Fig. 4.1.

The selection for a particular type depends on the factors like process temperature and pressure, nature of the materials to be handled and products obtained, position of the vessel (horizontal or vertical), nature of the support and above all the economy. Uses of common types of heads are briefly narrated below.

Flat Heads : Flat covers are generally used for manholes in low pressure vessels or to blind any flanged opening. They can also be used as closures for small diameter vessels operating at low pressure. For larger vessels or at higher pressure, the flat head cover will be too bulky ; otherwise, it will tend to collapse. The discontinuity stresses at the junction of head and shell are high. From fabrication point of view, this is the simplest type head to construct just cutting a circular piece from a flat plate. As a result, for a particular diameter and operating conditions, material cost for flat head is maximum, though, fabrication cost is very low.

Flanged-only Heads : This is also a type of flat head in which the gradual change in the shape at the corner results in reduced local stresses. This head is very economical to fabricate and finds its widest application in closing the ends of horizontal storage vessels at atmospheric pressure to store liquids having low vapour pressures, such as, fuel oil, kerosene, etc. Also used as bottom heads of vertical cylindrical vessels that rest on concrete slab.

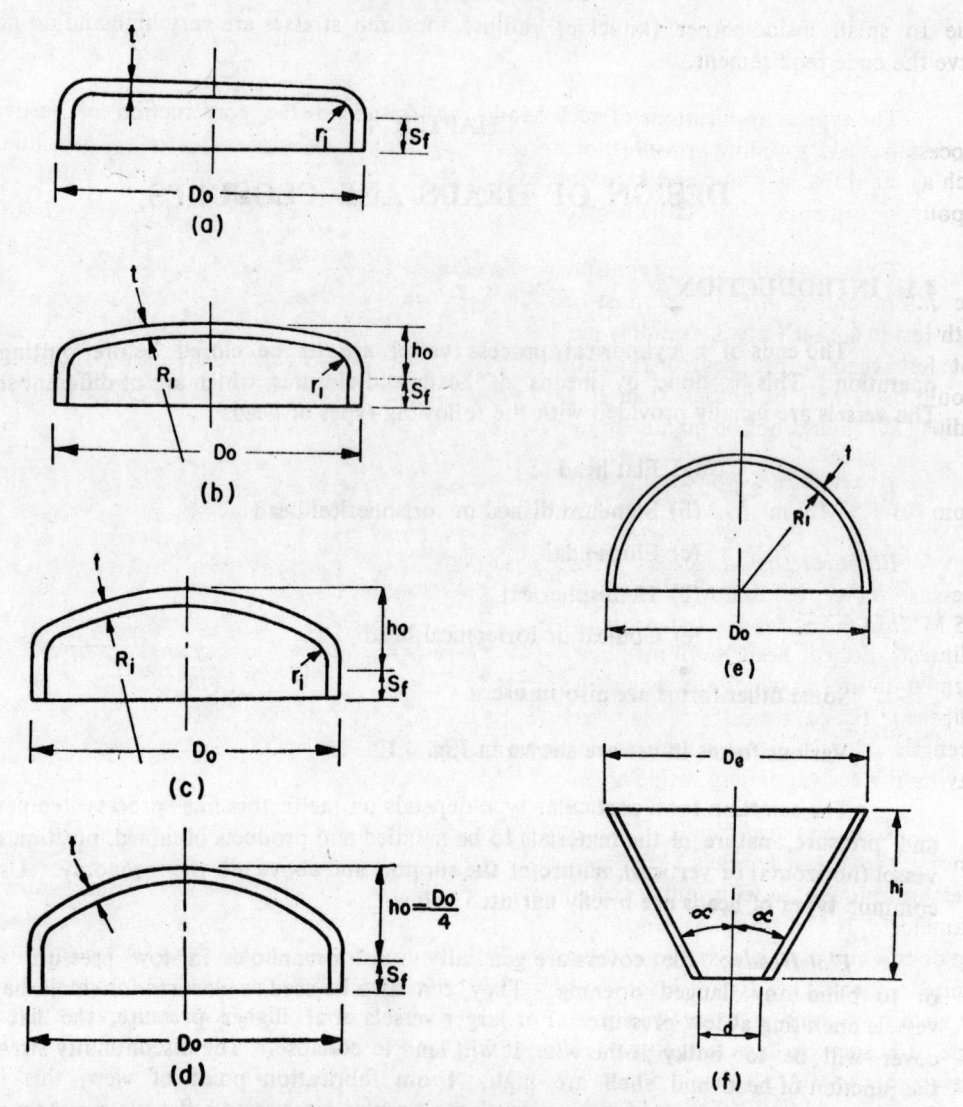

Fig. 4.1 Commonly used formed heads
(a) Flanged only, (b) Flanged and shallow dished, (c) Flanged and standard dished
   (torispherical), (d) Ellipsoidal, (e) Hemispherical, (f) Conical head or reducer.

Flanged Shallow Dished and Flanged Standard Dished Heads: While pressing into dished shape, such heads consist of two radii, namely, 'crown' radius and 'knuckle' radius. If the crown radius is greater than the shell outside diameter, the head is known as a 'flanged and shallow dished head'. On the other hand, if the crown radius is equal to or less than the outside diameter, the head is known as 'flanged and standard dished head'.

Due to small inside-corner (knuckle) radius, localized stresses are very high and do not serve the code requirement.

The typical applications of such heads are found in the construction of vertical process vessels for low pressures, of horizontal cylindrical storage tanks for volatile fluids, such as, naphtha, gasoline and kerosene, and of large diameter storage tanks in which the vapour pressure and hydrostatic pressure are too great for the 'flanged-only heads'.

To increase the pressure rating of flanged and dished heads, the local stresses at the inside corner of the head must be reduced. This can be achieved by forming the head with inside corner (knuckle) radius not less than three times the metal thickness and also not less than 6 per cent of the inside diameter. As per IS code inside corner radius (r_i) should preferably be not less than 10 per cent of inside diameter and also inside crown radius (R_i) should not be greater than outside diameter (D_o).

Heads of this type are used for pressure vessels in the general pressure range of from 0.1-1.5 MN/m².

Elliptical Dished Heads : These type of heads are generally recommended in the pressure range of 0.7 MN/m² and preferably for the vessels designed to operate above 1.5 MN/m², an economic balance may be made to select between standard dished head and elliptical dished head both meeting the IS code specifications. Elliptical dished heads are formed on dies in which the diametrical cross-section is an ellipse. Most of the standard elliptical dished heads are manufactured on 2 : 1 ratio of major to minor axis. The strength of such head is approximately equal to the strength of a seamless cylindrical shell having the corresponding inside and outside diameters.

Hemispherical Heads : For a given thickness, this type of heads is the strongest among the formed heads. These heads can be used to resist approximately twice the pressure rating of an ellipsoidal dished head or cylindrical shell of the same thickness and diameter. The degree of forming and the accompanying costs are greater than for any of the commonly used heads. Also the available sizes formed from single plates are more limited. Small heads are made by spinning, but large hemi-spherical heads are fabricated by welding pressed plate sections in the shape of a crown and petals, or by forging. This type of head is perhaps the most expensive but it is widely used in heavy duty high pressure vessels due to the fact that the most efficient use of the material is achieved.

Conical Heads and Reducers : Conical heads are used as bottom for a variety of process equipment like evaporators, spray driers, crystallizers, settling tanks, silos, etc. The particular advantage lies in the accumulation and removal of solids from such equipment. Another common application of conical head is as reducer, providing a smooth transition between two parts of different diameter in cylindrical process vessels. Conical heads with a sharp cylinder-cone junction is easiest to manufacture, but this simplicity is partly lost, when a knuckle radius is specified, to reduce the stress-concentration at the junction. When the apex angle is more than 60° or fatigue failure is expected, it is necessary to provide a transition knuckle radius. Cones having apex angle 60° are commonly used for the removal of solids.

Formed Heads Sections and their Significance : All formed heads are made of three integral sections. They are : (1) central dished section (2) corner torus section and (3) straight flange section. Central dished section is nothing but a part of a sphere having the radius of curvature equal to crown radius (R). The strength of this section is same as that of a spherical vessel with same radius of curvature. To convert the dished section smoothly into a cylindrical cross-section, corner torus is formed with radius of curvature called knuckle radius which is much smaller than the crown radius. As a result discontinuity occurs at the junction of torus and dished sections causing stress-concentration in this region. In designing formed head, stress consideration of torus is more important than that of the dished section which is much stronger. Next to torus comes straight flange, which is having purely a cylindrical cross-section. Though, head is complete with dished and torus sections from fabrication point of view, a straight flange section is also very very essential to keep the torus section, where stress-concentration would be large, away from any welded joint which is necessary to connect head with the cylindrical shell to avoid thermal stresses to be induced in the torus during welding. Straight flange behaves like a cylinder under pressure. So, strength of this section is same as that of a cylindrical vessel under similar conditions. In designing head therefore, consideration of stress-concentration at the torus section is an important criteria. Thickness of the torus section, however is the determining factor to select the head thickness. Two junctions of discontinuity are influencing the stress-concentration in the torus section. These are torus-sphere and torus-cylinder. Larger the knuckle radius, stronger is the torus section. General practice is to provide knuckle radius not less than three times the head thickness or 6 per cent of the inside diameter of the head, whichever is larger. Due to the smaller radius of curvature, torus section gets thinner during forming. This reduction of thickness depends upon the radius of curvature and plate thickness. For general purpose a reduction of 6 per cent may be considered reasonable, though it can go upto 10% (IS : 4049-1968). Exact magnitude should be obtained from the manufacturer's catalogue. As it is mentioned earlier, torus thickness is actually calculated to determine the head thickness. Design equation will give the minimum torus thickness required. In deciding the standard plate thickness with which head is to be formed, reduction of thickness in the torus section is to be considered appropriately.

When head and shell are to be connected through removable joint like bolted flange, straight flange section in the head is used for welding the ring flange, with the head. Purpose is again not to connect the ring flange with the torus section to avoid the further enhancement of the stress-concentration.

Many times it will be found that under the same operating conditions, requirement of head thickness is considerably larger than that of shell thickness. To provide a smooth joint between head and shell under such condition, straight flange section can be chamfered for gradual reduction of thickness, so that, the edge-thickness becomes equal to the shell thickness. This operation will not affect the strength of the head any way.

Standard length of straight flange is from 40-100 mm. Maximum length[1] may go upto 200 mm, if specially ordered. But this will incur extra charges. Minimum length of straight flange is 3 times the end thickness and this should not be less than 20 mm (IS : 4049—1968).

4.2 ANALYSIS AND DESIGN

4.2.1 Flat cover head

The stress analysis of flat cover head is made here on the basis of bending of uniformly loaded circular plates of constant thickness. For deriving the general expression for head thickness, two situations are considered. In one case, edges of the plate are assumed to be clamped preventing it from rotating only and not otherwise restrained, i.e. there is no strain in the neutral plane of the plate. In this case there are negative end bending moments.[2,3] In this second case, edges are considered to be freely supported, thus eliminating the edge bending moment.

For uniformly loaded circular plate, general deflection equation is given by[3,4]

$$w = \frac{p\,x^4}{64\,D} - C_1 \frac{x^2}{4} - C_2 \ln x + C_3 \qquad \ldots(4.2.1)$$

In the above equation,

p = load intensity (say pressure)

x = distance of any part of the plate under consideration from the centre,

D = flexural rigidity of the plate,

$\quad = \dfrac{E\,t^3}{12\,(1 - \mu^2)}$

t = thickness of the plate,

μ = Poisson's ratio,

E = modulus of elasticity,

C_1, C_2, C_3 = constants of integration,

Constants of integration are determined in each particular case of loading by the edge conditions of the plate.

---

[1] L. E. Brownell and E. H. Young, "Process Equipment Design", John Wiley and Sons, Inc., New York, 1959, p. 93.

[2] J. H. Harvey, "Pressure Vessel Design", East-West Press Pvt Ltd., New Delhi, 1969.

[3] S. Timoshenko and S. Woinowsky-Krieger, "Theory of Plates and Shells", McGraw-Hill Book Co. Inc., New York, 1959.

[4] S. Timoshenko, "Strength of Materials", Part II, D. Van Nostrand Co., Inc., Princeton, N.J., 1956.

(a) For circular plate with clamped edges, the equation of deflection (Eq. 4.2.1) becomes,[2]

$$w = \frac{p}{64 D} (R^2 - x^2)^2 \qquad ...(4.2.2)$$

where R is the radius of the plate at the point of support.

The maximum deflection is at the centre of the plate ($x = 0$) equal to

$$\delta = \frac{p R^4}{64 D} \qquad ...(4.2.3)$$

If M_r and M_θ are bending moments per unit length caused by pressure and M_r acts along cylindrical sections and M_θ along diametrical sections of the plate,[2] then,

$$M_r = \frac{p}{16} [R^2 (1 + \mu) - x^2 (3 + \mu)] \qquad ...(4.2.4)$$

$$M_\theta = \frac{p}{16} [R^2 (1 + \mu) - x^2 (1 + 3\mu)] \qquad ...(4.2.5)$$

Moments at the edge of the plate is obtained by substituting $x = R$. This gives the following expressions:

$$M_r = - \frac{pR^2}{8} \qquad ...(4.2.6)$$

$$M_\theta = - \frac{\mu p R^2}{8} \qquad ...(4.2.7)$$

Similarly, by substituting $x = 0$ in Eqs. 4.2.4 and 4.2.5, the moments at the centre are obtained as follows:

$$M_r = M_\theta = \frac{1 + \mu}{16} p R^2 \qquad ...(4.2.8)$$

From Eqs. 4.2.6, 4.2.7 and 4.2.8 it is evident that the expression for maximum moment is given by Eq. 4.2.6. This indicates that the maximum stress is at the edge of the plate and equal to

$$\sigma_{max} = \sigma_r = \frac{M_r}{t^2/6} = \frac{6}{t^2} \times \frac{p R^2}{8}$$

$$= \frac{3}{4} \frac{p R^2}{t^2} \qquad ...(4.2.8)$$

DESIGN OF HEADS AND CLOSURES

(b) When the edges of a uniformly loaded circular plate is simply supported, the deflection equation becomes[2] :

$$w = \frac{p(R^2 - x^2)}{64 D}\left[\frac{(5 + \mu) R^2}{1 + \mu} - x^2\right] \qquad \ldots(4.2.9)$$

At $x = 0$, maximum deflection at the centre becomes,

$$\delta = \frac{(5 + \mu) p R^4}{64 (1 + \mu) D} \qquad \ldots(4.2.10)$$

Bending moments equations are[2] :

$$M_r = \frac{p}{16} (3 + \mu) (R^2 - x^2) \qquad \ldots(4.2.11)$$

$$M_\theta = \frac{p}{16} [R^2 (3 + \mu) - x^2 (1 + 3\mu)] \qquad \ldots(4.2.12)$$

The maximum bending moment occurs at the centre where,

$$M_r = M_\theta = \frac{(3 + \mu) p R^2}{16} \qquad \ldots(4.2.13)$$

and corresponding maximum stress is

$$\sigma_{max} = \sigma_r = \sigma_\theta = \frac{M}{t^2/6} = \frac{6}{t^2} \frac{(3 + \mu) p R^2}{16}$$

$$= \frac{3 (3 + \mu)}{8} \frac{p R^2}{t^2} \qquad \ldots(4.2.14)$$

In Eqs. 4.2.8 and 4.2.14, if 'σ_{max}' is substituted by allowable stress 'f' of the material and 'R' is replaced by $D_e/2$, where D_e is the effective diameter of the flat head, a general expression for calculating the thickness of flat heads and covers is obtained. This is as follows :

$$t = C D_e \sqrt{p/f} \qquad \ldots(4.2.15)$$

where, C is a factor depending upon the method of attachment to shell.

Eq. 4.2.15 is given in IS : 2825—1969 for calculating the flat head thickness.

Following are the few cases where C values are indicated (IS : 2825—1969).

1. Flanged flat heads butt welded to shell. $D_e = D_i$; $C = 0.45$.
2. Plates welded to the inside of the shell $D_e = D_i$; $C \geqslant 0.55$ (Fig. 4.2 a).
3. Plates welded to the end of the shell (no inside welding). $D_e = D_i$; $C = 0.7$. (Fig. 4.2 b)
4. Plates welded to the end of the shell with an additional fillet weld on the inside. $D_e = D_i$; $C = 0.55$. (Fig. 4.2 c)
5. Covers riveted or bolted with a full face gaskets to shells, flanges or side plates. D_e = Bolt-circle diameter ; $C = 0.42$. (Fig. 4.2 d)
6. Covers with a narrow face bolted flange joint, i.e., gasket is placed within bolt holes. D_e = mean diameter of gasket ;

$C = \left(0.31 + 190 \dfrac{F_B h_G}{p D_e^3}\right)^{\frac{1}{2}}$, where F_B is the bolt load and $h_G = \frac{1}{2}$ (bolt circle diameter $- D_e$). (Fig. 4.2 e)

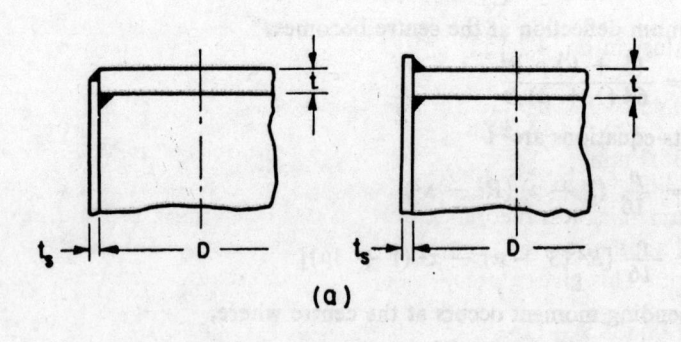

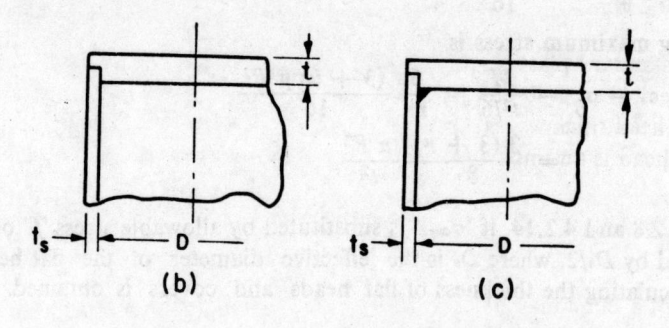

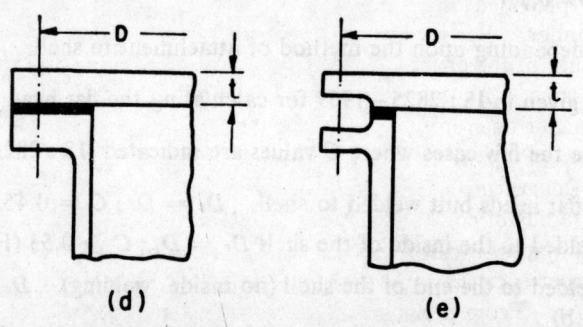

Fig. 4.2 Typical flat heads indicating the values of 'D'

Design Example 4.1 : Determine the thickness of the flanged flat covers of the process vessel of Example 3.1.

Solution : To determine the thickness of flanged flat head, Eq. 4.2.15 is to be used.
$$t = C D_e \sqrt{p/f} \qquad \ldots(4.2.15)$$

DESIGN OF HEADS AND CLOSURES

Assuming that the head is butt-welded to the shell, $C = 0.45$; $D_e = D_i = D_o - 2 \times t_s$. From the solution of Example 3.1, $D_i = 1\,190$ mm $= D_e$; $p =$ design pressure $= 0.525$ MN/m² ; $f = 118$ MN/m² (assuming that the material of construction for both head and shell is same).

Substituting the values in Eq. 4.2.15,

$$t = 0.45 \times 1\,190 \times (0.525/118)^{\frac{1}{2}} \text{ mm}$$

$$= 35.9 \text{ mm}$$

This is the minimum thickness theoretically calculated. To this 2 mm corrosion allowance is to be added and another say 6% is to be added to take care of the reduction in thickness at the torus section. This gives a practically required minimum thickness of

$$t = 1.06\,(35.9 + 2)$$

$$= 40.2 \text{ mm}$$

This necessitates a standard plate thickness 45 mm for the head. If reduction of thickness during forming is less, then a 40 mm plate can be used. It is to be noted, when shell thickness is only 6 mm, head thickness is 45 mm. It only indicates that the selection of the proper type of the head is an important criteria in process vessel design.

4.2.2 Conical head

Geometry of a cone may be compared with that of a cylinder in which the diameter is continuously changing. This suggests that the cylindrical shell equations may also be applied in this case, provided the diameter is appropriately modified incorporating the conical shape for a cylinder. For a cone, the radius of curvature at any section is not same as the radius of curvature of the element at that section in the hoop direction, while for a true cylinder both are same. As twice the radius of curvature of the element in the hoop direction gives the diameter term in the hoop stress equation, a relationship between radius of a cone and the radius of curvature in the hoop direction is necessary. In Fig. 4.3(a), R represents the radius of curvature of the element in the hoop direction (i.e. perpendicular to the meridian), r is the radius of cone at the same reference point, and α is half the apex angle. Then, by trigonometry it can be shown that $R = r/\cos \alpha$. This only implies that a cone of radius r will experience the same stress effect as will be observed in an equivalent cylinder of radius $r/\cos \alpha$. From this it may be concluded, if D_k is the inside diameter of cone at the position under consideration a relationship for wall thickness of cone will be obtained by substituting $D_k/\cos \alpha$ for D_i in Eq. 3.3.19, i.e.,

$$t = \frac{p\,D_k}{2\,fJ - p} \times \frac{1}{\cos \alpha} \qquad \ldots(4.2.16)$$

Where, $t =$ thickness of the cone,

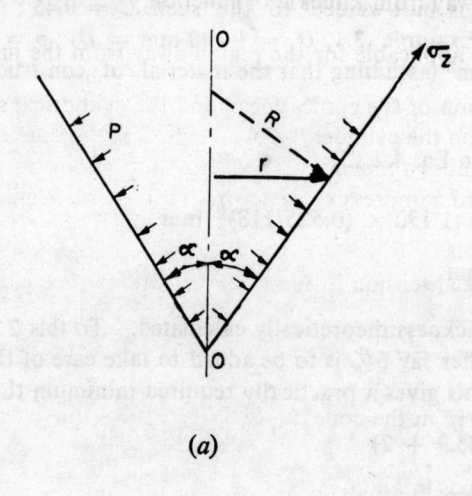

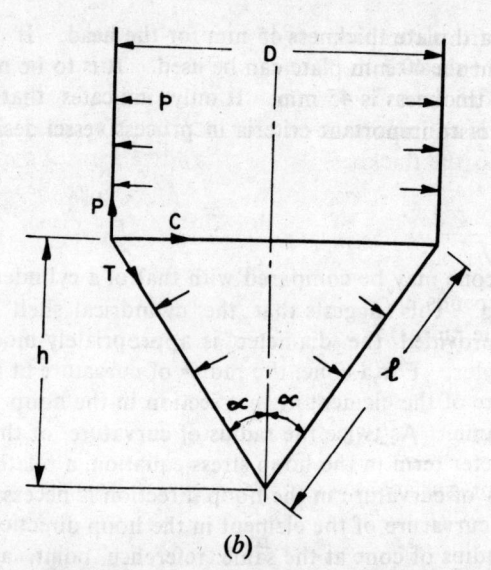

Fig. 4.3 Stresses in a conical head

From the consideration of stress analysis, a cone is divided into two regions. They are:

(a) region around knuckle or junction not exceeding a distance, $\frac{1}{2}(D_e\, t/\cos \alpha)^{\frac{1}{2}}$, from the junction or knuckle,[5] where D_e is the outer diameter of the conical section or end;

---
[5] IS : 2825—1969, Indian Standards Institution, New Delhi.

DESIGN OF HEADS AND CLOSURES

(b) region away from knuckle or junction.

Eq. 4.2.16 is applicable for the region away from the junction.

At the junction of the conical head and the cylindrical shell, a compressive force is exerted by the cone on the cylinder. Due to internal pressure shell tries to expand radially outward and this results in a bending moment and shear at the junction. Referring to Fig. 4.3(b), the inward compressive force produced by the conical head can be evaluated as follows.

If P is the axial tension in the shell per unit of shell circumference, then

$$\pi D P = \frac{\pi D^2}{4} p \quad \text{or} \quad P = \frac{pD}{4}$$

If T is the axial tension in the cone due to internal pressure, at equilibrium,

$$P = T \cos \alpha, \text{ and } C = T \sin \alpha = P \tan \alpha = \frac{pD}{4} \tan \alpha.$$

As a result of this compressive force, it is impossible to design a conical head to eliminate moment and shear at the junction, since cone always tends to deflect inward, and the shell outward under the influence of internal pressure. A factor (Z) taking into account the influence of this discontinuity stresses is therefore introduced into the basic thin-shell equation to determine the thickness of the cone at the junction or knuckle. The equation is as follows[5]:

$$t = \frac{p D_e Z}{2 f J} \qquad \qquad \ldots(4.2.17)$$

For single cone Z depends upon apex angle and knuckle radius. Z values for single sharp cone (without knuckle radius) are given below.[5]

$\alpha =$	20°	30°	45°	60°
$Z =$	1.00	1.35	2.05	3.2

Equations for surface area and volumetric capacity of a conical head are given below:

$$A = \tfrac{1}{2} \times \pi D \times l$$

$$= 1.57 D \sqrt{\frac{D^2}{4} + h^2} \qquad \qquad \ldots(4.2.18)$$

$$V = \tfrac{1}{3} \times \frac{\pi D^2}{4} \times h$$

$$= 0.131 D^3 / \tan \alpha \qquad \qquad \ldots(4.2.19)$$

Design Example 4.2 : If the process vessel of Example 3.1 is to be provided with a conical head having the half-apex angle 30°, determine the thickness of the head.

Solution : (a) Thickness of the head at the junction of the head and shell is to be evaluated from Eq. 4.2.17.

$$t = \frac{pD_oZ}{2fJ}$$

where
$p = 0.525$ MN/m^2
$D_o = 1.2$ m
$f = 118$ MN/m^2
$J = 0.85$
$Z = 1.35$

Substituting,

$$t = \frac{0.525 \times 1.2 \times 1.35}{2 \times 118 \times 0.85} = 4.24 \times 10^{-3} \text{ m}$$

$$= 4.24 \text{ mm}$$

(b) Thickness away from the junction is determined from Eq. 4.2.16.

$$t = \frac{pD_k}{2fJ - p} \times \frac{1}{\cos \alpha}$$

Here D_k is the maximum inside diameter of the cone at a distance $\frac{1}{2}(D_o\, t/\cos \alpha)^{\frac{1}{2}}$ from the junction.

$$\tfrac{1}{2}\left(\frac{D_o\, t}{\cos \alpha}\right)^{\frac{1}{2}} = \tfrac{1}{2}\left(\frac{1\,200 \times 7}{0.866}\right)^{\frac{1}{2}}$$

$$= 50 \text{ mm}$$

t is taken 7 mm including corrosion allowance. If inner diameter of the cone at the junction is same as that of the shell, i.e. 1 188 mm, then from trigonometry it can be seen that at a slant distance of 50 mm from the junction $D_k = 1\,188 - 2 \times 25 = 1\,138$ mm. Substituting the values,

$$t = \frac{0.525 \times 1\,138}{2 \times 118 \times 0.85 - 0.525} \times \frac{1}{0.866} \text{ mm}$$

$$= 3.45 \text{ mm}$$

As the difference in thickness comes less than 1 mm in two regions of the cone, there is no justification in using different thickness sheets for two sections. It will be rather economical from fabricational point if single thickness is used. In such case, the larger of the two is to be chosen. If the corrosion allowance of 2 mm is added to the minimum calculated thickness of 4.24 mm, which is required, a standard sheet metal thickness 7 mm will be required for the conical head. It is interesting to observe that for the same pressure vessel, a flanged flat head thickness of 45 mm is required (see Example 4.1).

4.2.3. Torispherical (standard dished) and ellipsoidal dished heads

Any dished head consists of 3 integral parts, namely, central dishing, corner torus and end straight flange (Fig. 4.4). All the 3 parts are having different meridional radius of curvature (meridional radius is perpendicular to the direction of longitudinal stress). As a result there are two junctions of discontinuity existing in a formed head. First one is between knuckle (torus) and crown (dish) and second one between knuckle and straight flange.

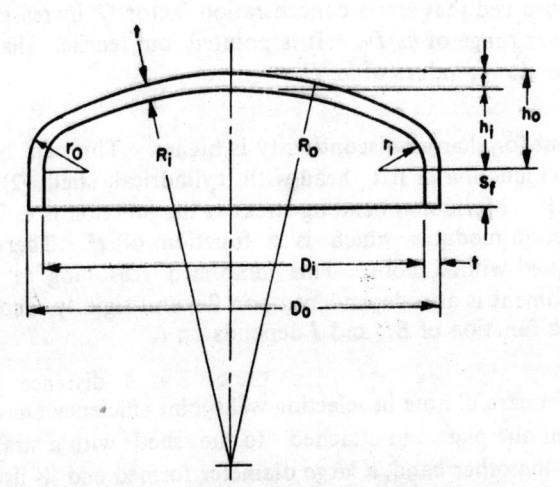

Fig. 4.4 Construction of a dished head

Due to differential dilations at the junctions under pressure loading, bending moments and shear stresses are induced, besides normal membrane stresses, to maintain the continuity between adjacent parts. As the torus section is small, these stress concentration factors cannot be neglected in the design equation. Coates[6] has given a relationship for local bending stresses at the junction of head and shell. According to the relationship suggested by C. O. Rhys,[7] the maximum bending stress is the meridional bending stress in the knuckle due to discontinuity at the crown-knuckle junction. Introducing a shape or stress

---

[6] W. M. Coates, "State of Stress in Full Heads of Pressure Vessels", Trans. Am. Soc. Mech. Engrs., 52 (1930), p. APM-52-12.
[7] C. O. Rhys, "Stresses in Dished Heads of Pressure Vessels", Trans. Am. Soc. Mech. Engrs., 53 (1931), p. PME-53-7.

concentration factor C, IS has given the following simplified relationship to evaluate the head thickness.

$$t = \frac{pD_oC}{2fJ} \qquad \ldots(4.2.20)$$

Here the shape factor C depends on h_E/D_o and t/D_o for the head without any opening or openings completely reinforced. In the case of heads containing unreinforced opening, C will depend on h_E/D_o and $\dfrac{d}{\sqrt{tD_o}}$. h_E is the effective external height of the head without straight flange and equal to the least of h_o or $D_o^2/4R_o$ or $(D_o r_o/2)^{1/2}$. It may be seen from Table 4.1, as h_E/D_o increases, C decreases. Reason can be attributed to the fact that, with the increase of h_E/D_o the difference in radius of curvature in torus and dished sections diminishes, thus reducing the discontinuity stresses at the knuckle-crown junction. The limiting value of h_E/D_o is 0.5 when the head becomes hemispherical. Again, from the same table it may be observed that stress concentration factor C increases sharply with the decrease of t/D_o for lower range of h_E/D_o. It is pointed out earlier that at the junction, discontinuity is more for lower values of h_E/D_o.

Bending moment for sharper discontinuity is higher. This can be proved from the following two cases—(1) junction of flat head with cylindrical shell, (2) junction of hemispherical head with shell. Meridonal bending stress at the junction is estimated by dividing bending moment by section modulus which is a function of t^2. Therefore, if t is small, i.e., t/D_o is less, C required will be more. This simplified reasoning is however, not sufficient, as the bending moment is also dependent upon flexural rigidity and damping factor.[2] These are expressed as a function of EI, and I depends on t.

One is to make a careful note in selecting weld-joint efficiency factor J for the heads. If the head is made from one plate and attached to the shell with a straight flange, J is to be taken as unity. On the other hand, a large diameter formed end is usually fabricated by welding petals with a central dish. In such cases and also in the case of circumferential end-to-shell seam having no straight flange, appropriate weld-joint factor J is to be chosen considering the nature of the weld and whether the seam is radiographed.

In $\dfrac{d}{\sqrt{tD_o}}$, d is the diameter of the largest uncompensated opening in the head. Design procedure for the compensation of or reinforcing the openings will be discussed in Ch. 6.

Eq. 4.2.20 is applied to determine the thickness of the formed ends, when the pressure is acting at the concave side. To calculate the thickness of ends convex to pressure, care is to be taken to ensure safety against elastic deformation. Eq. 4.2.20 can be applied

.or the second situation by increasing the design pressure p by 20 per cent.[5] The following equation gives the pressure p_E at which elastic deformation for the ends convex to pressure takes place[8].

$$p_E = 0.366\, E \left(\frac{t}{R_i}\right)^2 \qquad \text{...(4.2.21)}$$

The safety against elastic failure is ensured, if $p_E > 3p$.

The external height, h_o of a dished head, the approximate blank diameter of the plate with which formed ends can be fabricated and volume contained within heads are evaluated from the following expressions.[1,5]

h_o (excluding straight flange)

$$= R_o - \left[\left(R_o - \frac{D_o}{2}\right) \times \left(R_o + \frac{D_o}{2} - 2r_o\right)\right]^{\frac{1}{2}} \qquad \text{...(4.2.22)}$$

$$\text{Blank diameter} = D_o + \frac{D_o}{42} + \frac{2}{3}r_i + 2\,S_f \text{ (for } t \leqslant 25 \text{ mm)} \\
= D_o + \frac{D_o}{42} + \frac{2}{3}r_i + 2\,S_f + t \text{ (for } t > 25 \text{ mm)} \qquad \text{...(4.2.23)}$$

$$V \text{ (excluding straight flange)} = 0.084\,7\,D_i^3 \text{ (for } r_i = 0.06\,D_i) \\
\text{and } = 0.131\,3\,D_i^3 \text{ (for 2 : 1 ellipsoidal or deep dished head)} \qquad \text{...(4.2.24)}$$

In the case of hemispherical heads, the bending stresses due to discontinuity at the junction with cylindrical shell are small. The discontinuity stresses do not exceed the maximum membrane stresses by more than 4%, provided the cylindrical shell thickness to hemispherical shell thickness ratio is in the range of 0.8—2.0. Eq. 4.2.20 applies to hemispherical head.

Table 4.1(A) Stress Concentration Factor C for Formed Heads Without Opening or With Fully Compensated Openings[5]

	t/D_o				
h_E/D_o	0.002	0.005	0.01	0.02	0.04
0.15	4.55	2.66	2.15	1.95	1.75
0.20	2.30	1.70	1.45	1.37	1.32
0.25	1.38	1.14	1.00	1.00	1.00
0.30	0.92	0.77	0.77	0.77	0.77
0.40	0.59	0.59	0.59	0.59	0.59
0.50	0.55	0.55	0.55	0.55	0.55

[8] AD-MERKBLAETTER, Beuth-Vertrieb GMBH, Berlin.

Table 4.1(B) Stress Concentration Factor C for Formed Heads with Uncompensated Opening[5]

h_E/D_o	\multicolumn{6}{c}{$d/\sqrt{t\,D_o}$}
	0.5	1.0	2.0	3.0	4.0	5.0
0.15	1.67	1.86	2.15	2.65	3.10	3.60
0.20	1.28	1.45	1.85	2.30	2.75	3.25
0.25	1.00	1.15	1.60	2.05	2.50	2.95
0.30	0.83	1.00	1.45	1.88	2.28	2.70
0.50	0.60	0.80	1.10	1.50	1.85	2.15

Note : Values can be interpolated.

Design Example 4.3 : Determine the thickness of the plate required to fabricate a standard dished head for the process vessel of Example 3.1. It may be assumed that there will be no uncompensated opening in the head. Specifications for the head are given below :

$$R_i = D_o;\ r_i = 0.06\,D_o\ ;\ S_f = 40\ \text{mm}$$

What should be the blank diameter (i.e. the diameter of the plate) for the head ?

Solution. From Example 3.1 following data are known :

$p = 0.525\ \text{MN/m}^2$; $f = 118\ \text{MN/m}^2$; $D_o = 1.2$ m. Thickness of the head is determined from Eq. 4.2.20.

$$t = \frac{p\,D_o\,C}{2fJ} \qquad \ldots(4.2.20)$$

Solution of the above equation will require iteration, as C is also a function of t. As the first approximation assume $R_i = R_o = D_o$. Then from Eq. 4.2.22.

$$h_o = R_o - [(R_o - D_o/2) \times (R_o + D_o/2 - 2\,r_o)]^{\frac{1}{2}}$$

Substituting $R_o = D_o = 1.2$ m ; $r_o = 0.06 \times 1.2 = 0.072$ m

$h_o = 1.2 - 0.997 = 0.203$ m

$D_o^2/4\,R_o = 1.2/4 = 0.3$ m

$$\left(\frac{D_o\,r_o}{2}\right)^{\frac{1}{2}} = \left(\frac{1.2 \times 0.072}{2}\right)^{\frac{1}{2}} = 0.209$$

DESIGN OF HEADS AND CLOSURES

Out of the three quantities calculated above, h_o is the least. Therefore, the effective external height of the head is equal to h_o i.e. $h_E = 0.203$ m. From this,

$$h_E/D_o = \frac{0.203}{1.2} = 0.17$$

As the diameter of the vessel (1.2 m) is not very large head can be fabricated from a single plate and therefore, $J = 1$.

From Eq. 4.2.20,

$$t/D_o = \frac{pC}{2fJ} = \frac{0.525\ C}{2 \times 118 \times 1} = 2.225 \times 10^{-3}\ C$$

With the help of Table 4.1 (A), various values of $\frac{t}{D_o C}$ are to be tried for $h_E/D_o = 0.17$ to match the above relationship. It is found that for $t/D_o = 0.005$ and corresponding value of $C = 2.26$ gives $t/D_o\ C = 2.215 \times 10^{-3}$ which is a very good approximation. An exact value may be obtained by plotting $\frac{t}{D_o C}$ against C. From the above finding, the corroded head thickness is obtained as follows:

$$\frac{t}{D_o} = 0.005\ \text{or}\ t = 0.005 \times 1.2$$

$$= 6.0 \times 10^{-3}\ \text{m}$$

If corrosion allowance, 2 mm, is added to that, the required uncorroded plate thickness will be 8 mm. It is to be noted here that for the thinning of the torus no extra allowance is provided. If 6% allowance is given, then a standard plate thickness of 9 mm will be required. Compared to 1.2 m, 9 mm is very small. Hence, the first assumption of $R_i = R_o$ does not introduce any considerable error in the result. However, for accuracy it is suggested to recalculate h_o by putting $R_o = R_i + t = (1.2 + 9 \times 10^{-3})$ m and then calculate t from new value of h_E/D_o. Another method would be to assume some value for t and check the same from Eq. 4.2.20.

Blank diameter of the plate for forming the head can be evaluated from Eq. 4.2.23.

$$\text{Blank diameter} = D_o + \frac{D_o}{42} + \frac{2}{3}\ r_i + 2\ S_f$$

$$= 1\ 200 + \frac{1\ 200}{42} + \frac{2}{3}\ (72 - 9) + 2 \times 40$$

$$= 1\ 352\ \text{mm}$$

Design Example 4.4: For the process vessel of Example 3.1, a 10 mm thick 2:1 ellipsoidal head with an uncompensated opening is to be fabricated. What maximum diameter opening will be permissible?

Solution: Neglecting thinning effect, the corroded thickness $t = 10 - 2 = 8$ mm

From Eq. 4.2.20

$$C = \frac{2fJt}{pD_o}$$

Where,
$f = 118$ MN/m^2
$J = 1.0$
$t = 8 \times 10^{-3}$ m
$p = 0.525$ MN/m^2
$D_o = 1.2$ m

Substituting the values,

$$C = \frac{2 \times 118 \times 1 \times 8 \times 10^{-3}}{0.525 \times 1.2}$$

$$= 3.0$$

For 2:1 ellipsoidal head, $h_E = h_o = 0.25 D_o$,

Therefore, $h_E/D_o = 0.25$

Now, from Table 4.1 (B), for the magnitudes of $h_E/D_o = 0.25$ and $C = 3.0$, $d/\sqrt{tD_o}$ is to be obtained by interpolation. This gives

$$d/\sqrt{tD_o} = 5.11$$

or
$$d = 5.11 \times (tD_o)^{\frac{1}{2}}$$

$$= 5.11 \times (8 \times 1200)^{\frac{1}{2}} \text{ mm}$$

$$= 500 \text{ mm}$$

For a vessel diameter of 1.2 m, an opening diameter of 500 mm is quite large. However, because of the thickness much more than required is provided, such big opening is arrived at. Actual corroded thickness requirement may be less than 4 mm.

CHAPTER 5

LOCAL STRESSES IN PROCESS EQUIPMENT DUE TO DISCONTINUITY

5.1 INTRODUCTION

In process equipment the shape and dimensions of various parts are usually not same. This results in discontinuity stresses at the junctures due to unequal growth of the various parts subjected to pressure. It is seen in section 2.11, the dilation or radial growth of a sphere is different from that of a cylinder. If a hemispherical head is, therefore, welded to a cylindrical shell and subjected to an internal pressure, a discontinuity in the shape at the junction resulting from the differential growth or dilation will occur. But due to weldment or other kind of joint a continuity at the junction of two dissimilar parts is achieved under the influence of induced shear load P_o and moment M_o (Fig. 5.1), which in turn cause stresses in and around the juncture. Fortunately these are

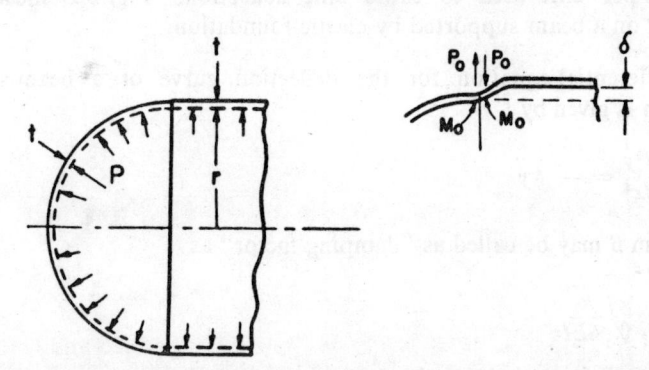

Fig. 5.1 Discontinuity at the junction of cylindrical shell and hemispherical head

local stresses and secondary in nature. The secondary stresses are those stresses developed by the constraint of adjacent parts. The basic characteristic is that they are self limiting. Local yielding or minor distortion can satisfy the conditions causing stress to occur and failure is not expected in one application. For example, stress concentration in a process vessel subjected to only a steady pressure or one repeated for only a few times, are of little importance provided the vessel is made of a ductile material like mild steel, which yields at these highly stressed locations allowing the stress to be transferred from overstressed fibres to the adjacent understressed ones. On the other hand, when the loading is a repetitive one, localized stresses become significant even though the material is ductile shortening its fatigue life. Therefore, the stresses given by the ordinary formulae which are based on average stress conditions, and which do not account for such local effects, must be multiplied by a theoretical "stress concentration factor" K_t defined as the ratio of the maximum stress to the average stress to obtain the maximum stress value.

To evaluate the extent and magnitude of such localized stresses occurring at shell-to-head junctions; support skirt, bimetallic joint, ring flange to nozzle or shell, etc., the concept of the behaviour of a "beam on elastic foundation" is applied. [1,2]

5.2 TERMINOLOGY

5.2.1 Behaviour of a beam on an elastic foundation

A practical example of, how a beam on an elastic foundation may behave can be seen by the nature of the deflection of a rail as the carriage passes over it. From this observation it can be stated, if a beam rests on an elastic foundation and is subjected to a concentric load P, the beam will deflect producing a continuous distributed reaction force q in the foundation proportional to the deflection y of the beam at the corresponding point; i.e. $q = ky$ per unit length. This force opposes the deflection of the beam. The proportionality constant k is called "spring constant" or foundation modulus, and is equal to the force required per unit area to cause unit deflection. Fig. 5.2 shows the effect of concentric load P on a beam supported by elastic foundation.

The differential equation for the deflection curve of a beam supported on an elastic foundation is given by [3]

$$EI \cdot \frac{d^4 y}{dx^4} = - ky \qquad \ldots (5.2.1)$$

Introducing a term β may be called as "damping factor" as

$$\beta = \sqrt[4]{\frac{k}{4EI}} \qquad \ldots (5.2.2)$$

the general solution of Eq. 5.2.1 is given by, [3]

$$y = e^{\beta x}(C_1 \cos \beta x + C_2 \sin \beta x) + e^{-\beta x}(C_3 \cos \beta x + C_4 \sin \beta x) \qquad \ldots (5.2.3)$$

Here y is the deflection at any distance x from the origin. The constants of integration C_1, C_2, C_3 and C_4 are to be determined from the known physical conditions. Depending upon the size and loading conditions, a process equipment can either be compared with infinitely long beam or semi-infinite beam for all practical purposes.

---

[1] M. Hetenyi, "Beams on Elastic Foundation", The University of Michigan Press, Ann Arbor, 1958.

[2] John F. Harvey, "Pressure Vessel Design", East-West Press Pvt. Ltd., New Delhi, 1969.

[3] H. B. Phillips, "Differential Equations", John Wiley & Sons, Inc., New York, 1934.

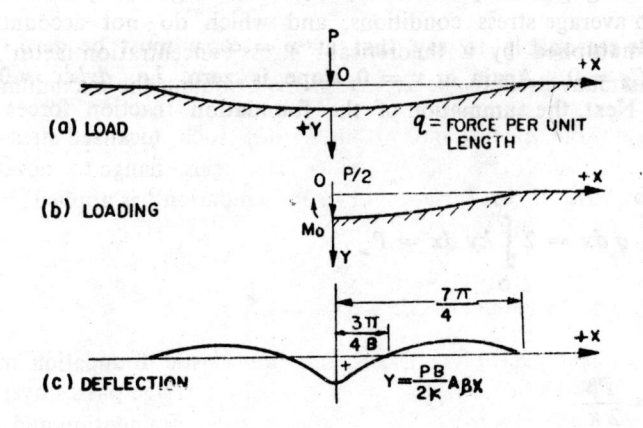

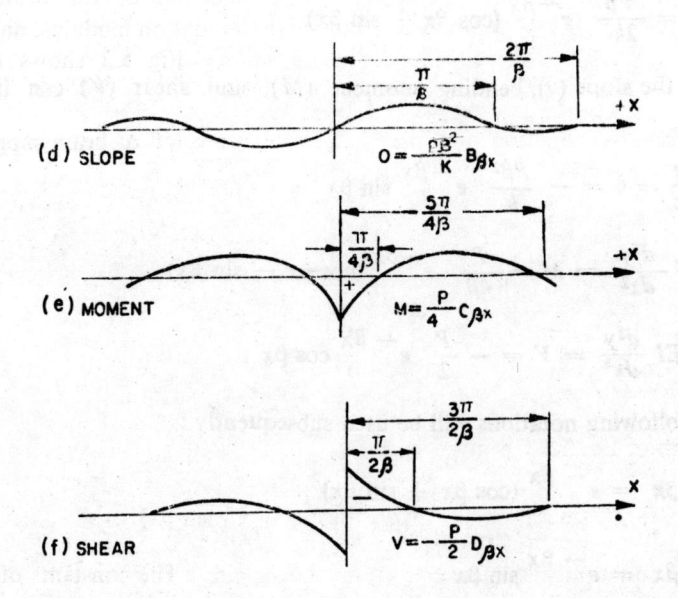

LOCAL STRESSES IN PROCESSES EQUIPMENT DUE TO DISCONTINUITY

(a) LOAD q = FORCE PER UNIT LENGTH

(b) LOADING

(c) DEFLECTION $Y = \dfrac{P\beta}{2K} A_{\beta x}$

(d) SLOPE $\theta = -\dfrac{P\beta^2}{K} B_{\beta x}$

(e) MOMENT $M = \dfrac{P}{4\beta} C_{\beta x}$

(f) SHEAR $V = -\dfrac{P}{2} D_{\beta x}$

Fig. 5.2 Effect of concentric load P on a beam supported by elastic foundation

5.2.2 Infinitely long beam subjected to a concentric load P (Fig. 5.2)

It is quite reasonable to say that at $x = \infty$, y must be zero. That means in Eq. 5.2.3 $C_1 = C_2 = 0$. Again, at $x = 0$, slope is zero, i.e. $dy/dx = 0$ and this gives $C_3 = C_4 = C$. Next, the summation of the foundation reaction forces must equal the applied force, i.e.

$$2\int_0^\infty q\,dx = 2\int_0^\infty ky\,dx = P$$

and from this,

$$C = \frac{P\beta}{2k}$$

Eq. 5.2.3 for infinitely long beam, therefore becomes,

$$y = \frac{P\beta}{2k} e^{-\beta x}(\cos\beta x + \sin\beta x) \qquad \ldots (5.2.4)$$

Expressions for the slope (θ), bending moment (M), and shear (V) can be found from Eq. 5.2.4

$$\frac{dy}{dx} = \theta = -\frac{P\beta^2}{k} e^{-\beta x} \sin\beta x \qquad \ldots (5.2.5)$$

$$EI\frac{d^2y}{dx^2} = M = \frac{P}{4\beta} e^{-\beta x}(\cos\beta x - \sin\beta x) \qquad \ldots (5.2.6)$$

$$-EI\frac{d^3y}{dx^3} = V = -\frac{P}{2} e^{-\beta x} \cos\beta x \qquad \ldots (5.2.7)$$

For simplicity following notations will be used subsequently:

$$A_{\beta x} = e^{-\beta x}(\cos\beta x + \sin\beta x)$$

$$B_{\beta x} = e^{-\beta x} \sin\beta x$$

$$C_{\beta x} = e^{-\beta x}(\cos\beta x - \sin\beta x)$$

$$D_{\beta x} = e^{-\beta x} \cos\beta x$$

The nature of the variation with distance from the point of applied load of the deflection, slope, moment and shear are shown in Fig. 5.2. The maximum values of deflection, moment and shear occur at the point of application of load, $x = 0$, and are found from the following equations.

$$y_{max} = \frac{P\beta}{2k} \quad \ldots (5.2.4a)$$

$$M_{max} = M_o = \frac{P}{4\beta} \quad \ldots (5.2.6a)$$

$$V_{max} = -\frac{P}{2} \quad \ldots (5.2.7a)$$

5.2.3 Semi-infinite beam

A semi-infinite beam has unlimited extension in one direction, but has also a finite end where load is acting. Force P will bend such a beam and reactive force will cause bending moment M_o at the origin. Eq. 5.2.3 is applicable here also. Applying the conditions at $x = \infty$, the equation of deflection for semi-infinite beam is given by,[2]

$$y = \frac{2P\beta}{k} D_{\beta x} - \frac{2M_o \beta^2}{k} C_{\beta x} \quad \ldots (5.2.8)$$

Similarly,

$$\theta = -\frac{2P\beta^2}{k} A_{\beta x} + \frac{4M_o \beta^3}{k} D_{\beta x} \quad \ldots (5.2.9)$$

$$M = -\frac{P}{\beta} B_{\beta x} + M_o A_{\beta x} \quad \ldots (5.2.10)$$

$$V = -PC_{\beta x} - 2 M_o \beta B_{\beta x} \quad \ldots (5.2.11)$$

The deflection and slope are maximum at $x = 0$, where,

$$y_{max} = \frac{2P\beta}{k} - \frac{2 M_o \beta^2}{k} \quad \ldots (5.2.8a)$$

$$\theta_{max} = -\frac{2 P \beta^2}{k} + \frac{4 M_o \beta^3}{k} \quad \ldots (5.2.9a)$$

For solving the problems involving discontinuity stresses at the junction of heads and shells, flange joints etc., the above equations, in conjunction with the principle of superimposition, can be applied.

5.2.4 Cylindrical vessels under axially symmetrical loading

Most of the process vessels, at the junction of dissimilar parts, are subject to rotationally symmetrical loading, but variable along its length. To such conditions the theory of beam on elastic foundations is ideally applied. Each longitudinal element of unit circumferential width can be considered acting as a 'beam' and remainder of the supporting cylinder as an "elastic foundation". The radial displacement due to loading can be considered as deflection of those longitudinal elements. The foundation modulus, k, for cylindrical vessels are given by,[2]

$$k = \frac{E t}{r^2} \qquad \qquad \ldots (5.2.12)$$

For cylindrical vessels the sides of each longitudinal element are not able to rotate to accommodate the lateral extension or compression resulting from the Poisson effect, as any change in the shape of cross-section of the longitudinal element is prevented by the adjacent elements of the cylinder. This causes bending moment in the circumferential section and is equal to

$$M_c = \mu\, M_x \qquad \qquad \ldots (5.2.13)$$

where M_x is the longitudinal bending moment and μ is Poisson's ratio. This is similar to the condition occurring in the flat plates.[2] This is taken into account by using "flexural rigidity" D. As the longitudinal elements are considered as beam of unit width, the following relationship is given.[4]

$$EI = D = \frac{E t^3}{12(1-\mu^2)} \qquad \qquad \ldots (5.2.14)$$

Substituting this value for EI and that for k from Eq. 5.2.12, the damping factor β for cylindrical vessel is obtained from Eq. 5.2.2 as,

$$\beta = \sqrt[4]{\frac{k}{4EI}} = \frac{\sqrt[4]{3(1-\mu^2)}}{\sqrt{rt}} \qquad \qquad \ldots (5.2.15)$$

For steel, $\mu = 0.3$, the above expression becomes

$$\beta = \frac{1.285}{\sqrt{rt}} \qquad \qquad \ldots (5.2.16)$$

From Eq. 5.2.14 it is seen that I of elastic beam is equal to I of the cantilever beam divided by $1 - \mu^2$.

5.2.5 Length of cylindrical vessels effected by load deformations

From Eqs. 5.2.4 through 5.2.7 and from Fig. 5.2 it is seen that the magnitude of the deflection, slope, bending moment, and shear all have the characteristic damped wave

---

[4] L. E. Brownell and E. H. Young, "Process Equipment Design", John Wiley & Sons, Inc., New York, 1959.

form of rapidly diminishing amplitude. The length of this wave is given by the period of the functions $\cos \beta x$ and $\sin \beta x$ and this is equal to :

$$2L = \frac{2\pi}{\beta} = 2\pi \sqrt[4]{\frac{4EI}{k}} \qquad \ldots (5.2.17)$$

It may be noted that the values of deflection etc. are all very small at about a distance $x = \pi/\beta$ on either side of the load. This indicates that a cylindrical steel vessel of length greater than $\pi \sqrt{rt} / 1.285$ or $2.45 \sqrt{rt}$ on either side of the load will act as if it were infinitely long and all infinitely long beam equations can be applied without any appreciable error.

From the above discussion it is seen that at a distance $x = \pi/\beta$ from the applied load, the effects of the load deformations are practically nil. This is one observation of elastic foundation characteristics. On the other hand, if one is to find the distance beyond the application of the load at which structural reinforcement of a pressure vessel may be assumed to have no significant effect, obviously it will not be at a distance $x = \frac{\pi}{\beta}$, but is somewhat closer to the point of load application. Harvey[2] has shown one simple approach to find the effective distance by approximating the elastic foundation deflection and slope characteristics with those of a similarly loaded equivalent length cantilever beam, which has similar deflection and slope curves, to give the same maximum values of these characteristics. The equivalent length cantilever beam is taken as the farthest distance from the point of application of the load at which a significant effect is registered. It is found by equating the end deflections and slopes of a beam on an elastic foundation loaded by a force P and moment M_o with those for a similarly loaded cantilever. From these comparisons it has been found that the average value is approximately $L = 1/\beta$. This approach gives a good "feel" of the elastic foundation beam behaviour and it establishes a practical limit for its primary effect, such as the reinforcing limits around vessel openings, minimum hub length for integrated flange, etc.

5.3 EVALUATION OF DISCONTINUITY STRESSES IN PROCESS VESSELS

In Ch. 3, direct tensile stresses caused by internal pressure was discussed. Membrane stresses were assumed uniform and no differential displacement of any part of the body due to varying magnitudes of membrane stresses was considered. But regional or local discontinuities in shape and size produce differential displacement in the vessel causing bending of the wall. These bending stresses, through local in extent, may become very high in magnitude. One such location would be at the juncture of the cylindrical shell with its closure head (Fig. 5.1) where the radial growth of the cylindrical portion of the vessel is not the same as that of the head when the vessel is pressurized. As a result, at the junction of these parts local bending takes place to preserve the continuity of the vessel wall. The additional stresses set up at these locations are called "discontinuity stresses".

In such cases the deformation and stress in the longitudinal or meridian elements are determined from the elastic foundation beam formulae. This is then added to the longitudinal pressure stress σ_a to obtain the combined or maximum longitudinal stress in that region. To determine the resulting stress in the circumferential direction, one is to add algebraically to the normal hoop pressure stress, σ_θ that due to direct compression (shortening) or tension (extension) of the radius giving σ_c, and that caused by circumferential bending σ_b. Thus the resulting circumferential stress in a cylindrical vessel is given by,

$$\sigma = \sigma_\theta \ (\pm) \ \sigma_c \ (\pm) \ \sigma_b \qquad \ldots (5.3.1)$$

Here, $\sigma_\theta = \dfrac{pr}{t}$ (for cylinder) ; $= \dfrac{pr}{2t}$ (for hemisphere)

$$\sigma_c = \dfrac{y_a}{r} E$$

$$\sigma_b = \dfrac{6}{t^2} \mu \ (M_a)$$

In the case of thin-walled vessels, t/r is small and therefore the deflection and bending becomes very local in extent and affects the stresses only in the immediate vicinity of the juncture. This narrow zone at the edge of the head can be considered as nearly cylindrical in shape. Therefore, the equations given for the cylindrical portions of the vessel can be used for approximate calculations of the deflections in spherical, elliptical, or conical shape head.[5]

5.3.1 Discontinuity stresses in cylindrical vessel with hemispherical head

Fig. 5.1 shows a cylindrical vessel with hemispherical head subjected to internal pressure. The expressions for hoop and longitudinal stresses are given in Ch. 3.

For cylindrical portion :

Hoop $\qquad \sigma_\theta = \dfrac{pr}{t} \qquad \ldots (5.3.2)$

Longitudinal $\quad \sigma_a = \dfrac{pr}{2t} \qquad \ldots (5.3.3)$

Radial growth $\quad \delta_c = \dfrac{pr^2}{2tE} (2 - \mu) \qquad \ldots (5.3.4)$

For spherical portion :

Hoop $\qquad \sigma_\theta = \dfrac{pr}{2t} \qquad \ldots (5.3.5)$

---

[5] J. P. Den Hartog, "Advanced Strength of Materials". McGraw-Hill Book Co. Inc., New York, 1952.

Longitudinal $\quad \sigma_s = \dfrac{pr}{2t}$ ... (5.3.6)

Radial growth $\quad \delta_s = \dfrac{Pr^2}{2tE}(1 - \mu)$ ... (5.3.7)

The difference in radial growth produced by membrane stresses in the two portions is given by (Fig. 5.1)

$$\delta = \delta_c - \delta_s = \dfrac{pr^2}{2tE} \quad ... (5.3.8)$$

Here thickness is assumed to be equal in both head and shell, which need not be the case always.

As shown in Fig. 5.1 the head and shell are kept together at the juncture by shearing forces P_0 and bending moments M_0 per unit length of circumference. These discontinuity forces produce local bending stresses in the adjacent parts of the vessel.

Case I: Thickness of the head and shell equal.

In this case the deflections and slopes induced at the edges of cylindrical and spherical parts, by the forces P_0 are equal. Therefore, the conditions of continuity at the juncture are satisfied if $M_0 = 0$ and P_0 is of such magnitude that it creates a deflection of $\delta/2$ at the edge of the cylinder. Here equations for semi-infinite beam are applicable. From Eq. 5.2.8 at $x = 0$ and $M_0 = 0$.

$$\delta/2 = \dfrac{2 P_0 \beta}{k} D_{\beta x} \quad ... (5.3.9)$$

Substituting k from Eq. 5.2.12, δ from Eq. 5.3.8 and noting $D_{\beta x} = 1$ at $x = 0$, P_0 is found as follows:

$$P_0 = \dfrac{p}{8\beta} \quad ... (5.3.10)$$

For any value of x the deflection and bending moment are obtained from Eqs. 5.2.8 and 5.2.10 respectively, when $P_0 = P$ is known.

Hence, the longitudinal stress at any point x from the point of juncture of cylinder and hemisphere is,

$$\sigma = \dfrac{pr}{2t} \pm \dfrac{6}{t^2} \mu \dfrac{P}{8\beta^2} B_{\beta x} \quad ... (5.3.11)$$

The resulting longitudinal stress in head and shell will be same. It is to be noted that $B_{\beta x} = 0$ at $x = 0$. The total hoop stress in the circumferential direction is given in Eq. 5.3.1. This will be different for head and shell for $x \neq 0$. Again, in case of cylinder

σ_c is negative as shortening radii causes compression. But P_0 produces an extension of radii for hemisphere, causing tension and thus σ_c is positive, for head. The total hoop stress for cylinder will be,

$$\sigma = \frac{pr}{t} - \frac{E}{r}\frac{p}{8\beta}\frac{2\beta}{k} D_{\beta x} \pm \frac{6}{t^2} \mu \frac{p}{8\beta^2} B_{\beta x} \qquad \ldots (5.3.12)$$

On substitution of $k = \frac{Et}{r^2}$ from Eq. 5.2.12, the above equation becomes,

$$\sigma = \frac{pr}{t} - \frac{pr}{4t} D_{\beta x} \pm \frac{3\mu p}{4 t^2 \beta^2} B_{\beta x} \qquad \ldots (5.3.13)$$

For hemispherical head total hoop stress is

$$\sigma = \frac{pr}{2t} + \frac{pr}{4t} D_{\beta x} \pm \frac{3\mu p}{4 t^2 \beta^2} B_{\beta x} \qquad \ldots (5.3.14)$$

At the junction, though, stresses in head and shell are same, it varies for $x \neq 0$.

Design Example 5.1: Determine the discontinuity stresses in the vessel at a distance $\beta x = \pi/4$ from the juncture of hemispherical head and cylindrical shell for the conditions given below.

$p = 2.0$ MN/m^2 ; $r = 1\,000$ mm ; $t = 20$ mm ; $\mu = 0.3$.

Solution : From Eq. 5.2.16

$$\beta = \frac{1.285}{\sqrt{rt}} = \frac{1.285}{\sqrt{1\,000 \times 20}} = 9.088 \times 10^{-3} \text{ mm}^{-1}$$

$$\beta^2 = 8.26 \times 10^{-5} \text{ mm}^{-2}$$

A. Longitudinal stress in the cylinder is given by Eq. 5.3.11 as

$$\sigma = \frac{pr}{2t} \pm \frac{6}{t^2} \mu \frac{p}{8\beta^2} B_{\beta x}$$

From section 5.2.2

$$B_{\beta x} = e^{-\beta x} \sin \beta x$$

$$= 0.322\,4 \text{ for } \beta x = \pi/4$$

Substituting for $B_{\beta x} = 0.322\,4$ in the above equation,

$$\sigma = \frac{2.0 \times 1\,000}{2 \times 20} \pm \frac{6}{(20)^2} (\cdot 3) \frac{2.0}{8 \times 8.26 \times 10^{-5}} \times 0.322\,4 \text{ MN/m}^2$$

$$= 50 \pm 4.39 \text{ MN/m}^2$$

The maximum value of longitudinal tensile stresses will be 54.39 MN/m²

B. Total hoop stress in the cylinder at $\beta x = \pi/4$ will be obtained from Eq. 5.3.13

$$\sigma = \frac{pr}{t} - \frac{pr}{4t} D_{\beta x} \pm \frac{3\mu p}{4 t^2 \beta^2} B_{\beta x}$$

From section 5.2.2,

$$D_{\beta x} = e^{-\beta x} \cos \beta x$$

$$= 0.3224$$

Substituting the values,

$$\sigma = \frac{2.0 \times 1\,000}{20} - \frac{2.0 \times 1\,000}{4 \times 20} \times 0.3224 \pm \frac{3 \times 0.3 \times 2.0}{4 \times 400 \times 8.26 \times 10^{-5}} \times 0.3224 \text{ MN/m}^2$$

$$= 100 - 8.06 \pm 4.39 \text{ MN/m}^2$$

The maximum hoop stress will be 96.33 MN/m² in the cylinder at $\beta x = \pi/4$.

Likewise, the total hoop stress in the hemispherical head is calculated from Eq. 5.3.14, which gives $\sigma = 62.45$ MN/m² which is a little less than the longitudinal tensile stress.

Case II : Thickness of head and shell not equal.

It is not uncommon in practice that hemispherical head is thinner than cylindrical shell. This is due to better membrane stress condition of the hemispherical head. From the foregoing sections it is seen that the discontinuity stresses are caused due to differential growth or dilation of the two adjacent parts of the vessel. In general it can be said, higher the value of differential growth, more will be the discontinuity stresses. That means, at zero differential growth no constraint is expected from the adjacent part and thus discontinuity stress becomes nil. It is also seen from Eqs. 5.3.4 and 5.3.7, the radial growth is inversely proportional to the thickness. From this a condition can be arrived, when the radial growth for both head and shell will be equal. If t_c is the thickness of the cylindrical shell and t_s is the thickness of the hemispherical head, then,

$$\delta_c = \frac{pr^2}{2 t_c E} (2 - \mu) \qquad \ldots (5.3.4)$$

$$\delta_s = \frac{pr^2}{2 t_s E} (1 - \mu) \qquad \ldots (5.3.7)$$

For $\delta_c - \delta_s = 0$,

$$\frac{t_s}{t_c} = \frac{1-\mu}{2-\mu} \qquad \ldots (5.3.15)$$

For $\mu = 0.3$, $\dfrac{t_s}{t_c} = \dfrac{1-0.3}{2-0.3} = 0.41$

From membrane stress consideration, t_s/t_c should not be less than 0.5. However t_s/t_c having any value larger than 0.41, differential growth and therefore, the discontinuity stresses cannot be avoided. As t_s/t_c increases, the discontinuity stresses also increase and for t_s/t_c equal to unity, the magnitudes of these stresses are determined earlier.

5.3.2 Discontinuity stresses in cylindrical vessel with flat head

Here head may be considered as a flat circular plate uniformly loaded by internal pressure p, and hence bends to a spherical surface with a corresponding change in slope at its juncture with the cylinder. This sets up a bending moment M_o which makes the cylindrical shell slope agree with the slope of the head. This satisfies one condition for continuity. The other condition requires that the radial growth of the cylinder be restricted at the head junction by P_o and M_o. Fig. 5.3 shows the force conditions at the junction. Subscripts h for head and c for cylinder are used. The continuity equations are as follows :

A. Slope continuity

$$\theta_{h,\,p} - \theta_{h,\,M_o} = \theta_{c,\,M_o} - \theta_{c,\,P_o} \qquad \ldots (5.3.16)$$

Here, $\quad \theta_{h,\,p}$ = head edge slope due to pressure p

$$= \frac{3\,pr^3}{2\,E\,t_h^3}(1-\mu) \qquad \ldots (5.3.17)$$

$\theta_{h,\,M_o}$ = due to edge moment M_o

$$= \frac{rM_o}{D(1+\mu)} \qquad \ldots (5.3.18)$$

$\theta_{c,\,M_o} - \theta_{c,\,P_o}$ = slope of cylindrical portion at the junction due to P_o and M_o

$$= \frac{4\,M_o\,\beta^3}{k}\,D_{\beta x} - \frac{2\,P_o\,\beta^2}{k}\,A_{\beta x} \qquad \ldots (5.3.19)$$

LOCAL STRESSES IN PROCESS EQUIPMENT DUE TO DISCONTINUITY

B. Radial displacement continuity

$$\delta_c = \delta_{h,\,Po} + \delta_{c,\,Po\,Mo} \qquad \ldots (5.3.20)$$

Here,
δ_c = unrestrained growth of the cylindrical portion

$$= \frac{pr^2}{2\,t_c\,E}\,(2 - \mu) \qquad \ldots (5.3.21)$$

$\delta_{h,\,Po}$ = radial deflection of head due to P_o

$$= \frac{rP_o}{t_h\,E}\,(1 - \mu) \qquad \ldots (5.3.22)$$

$\delta_{c,\,Po\,Mo}$ = radial deflection of cylindrical portion due to P_o and M_o

$$= \frac{2P_o\beta}{k}\,D_{\beta x} - \frac{2M_o\beta^2}{k}\,C_{\beta x} \qquad \ldots (5.3.23)$$

P_o and M_o are determined by solving Eqs. 5.3.16 and 5.3.20 simultaneously.

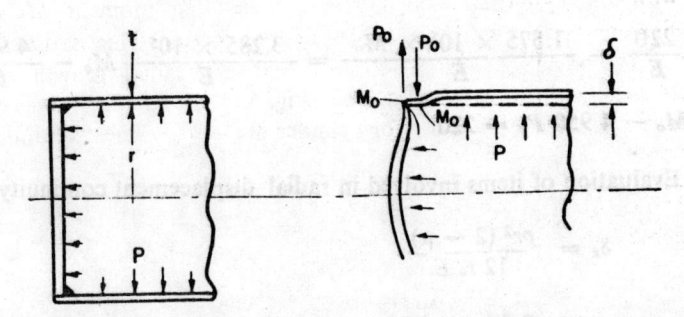

Fig. 5.3 Cylindrical shell with flat head and discontinuity at the junction

Design Example 5.2 : Determine the discontinuity stresses in the cylindrical shell at the junction with a flat head for the following conditions :

$p = 0.5$ MN/m²; $r = 0.15$ m; $t_h = 0.22$ m; $t_c = 0.01$ m;

$\mu = 0.3$ and $E = 1.6 \times 10^5$ MN/m².

Solution : (A) Evaluation of items involves in slope continuity :

$$\theta_{h,\,p} = \frac{3pr^3\,(1 - \mu)}{2\,E\,t_h^3}$$

$$= \frac{3 \times 0.5 \times (0.15)^3 \times 0.7}{2 \times (0.02)^3\,E} = \frac{220}{E}$$

70 CHEMICAL EQUIPMENT DESIGN—MECHANICAL ASPECTS

$$\theta_{h, M_o} = \frac{12\, r\, (1 - \mu^2)\, M_o}{E\, t_h^3\, (1 + \mu)}$$

$$= \frac{12 \times 0.15 \times 0.7 \times M_o}{8 \times 10^{-6}\, E} = \frac{1.575 \times 10^5}{E}\, M_o$$

$$\theta_{c,\, M_o} - \theta_{c,\, P_o} = \frac{4r^2\, \beta^3\, M_o}{E\, t_c} - \frac{2r^2\, \beta^2\, P_o}{E\, t_c}$$

$$= \frac{4 \times (0.15)^2 \times 36\,500}{0.01\, E}\, M_o - \frac{2 \times (0.15)^2 \times 1\,100}{0.01\, E}\, P_o$$

$$= \frac{3.285 \times 10^5}{E}\, M_o - \frac{4\,950}{E}\, P_o$$

$$\beta = \frac{1.285}{\sqrt{r\, t_c}} = \frac{1.285}{\sqrt{0.15 \times 0.01}} = 33.18\ m^{-1}$$

$$\beta^2 = (33.18)^2 = 1\,100\ ;\ \beta^3 = (33.18)^3 = 36\,500$$

Eq. 5.3.16 gives,

$$\frac{220}{E} - \frac{1.575 \times 10^5 \times M_o}{E} = \frac{3.285 \times 10^5}{E}\, M_o - \frac{4\,950}{E}\, P_o$$

or, $486\,000\, M_o - 4\,950\, P_o = 220$ ...(1)

(B) Evaluation of items involved in radial displacement continuity :

$$\delta_c = \frac{pr^2\, (2 - \mu)}{12\, t_c\, E}$$

$$= \frac{0.5 \times (0.15)^2 \times 1.7}{2 \times 0.01\, E} = \frac{0.956}{E}$$

$$\delta_{h,\, P_o} = \frac{r\, P_o\, (1 - \mu)}{t_h\, E}$$

$$= \frac{0.15 \times 0.7\, P_o}{0.02\, E} = \frac{5.25\, P_o}{E}$$

$$\delta_{c,\, P_o\, M_o} = \frac{2\, t^2\, \beta\, P_o}{E\, t_c} - \frac{2r^2\, \beta^2\, M_o}{E\, t_c}$$

$$= \frac{2 \times (0.15)^2 \times 33.18\, P_o}{0.01\, E} - \frac{2 \times (0.15)^2 \times 1\,100\, M_o}{0.01\, E}$$

$$= \frac{149.3\, P_o}{E} - \frac{4\,950\, M_o}{E}$$

Eq. 5.3.20 gives,

$$\frac{0.956}{E} = \frac{5.25 \, P_o}{E} + \frac{149.3 \, P_o}{E} - \frac{4\,950 \, M_o}{E} \qquad \ldots \text{(II)}$$

Solving (I) and (II)

$$P_o = 3.07 \times 10^{-2} \text{ MN per m of circumference}$$

$$M_o = 7.65 \times 10^{-4} \text{ MN m per m of circumference}$$

(a) Resulting longitudinal stress in the shell at the junction is given by,

$$\sigma = \sigma_z \pm \sigma_b$$

Where, σ_z = longitudinal pressure stress

$$= \frac{pr}{2\,t_c} = \frac{0.5 \times 0.15}{2 \times 0.01} = 3.75 \text{ MN/m}^2$$

σ_b = longitudinal bending stress due to P_o and M_o

$$= \frac{6\,M_o}{t_c^2} = \frac{6 \times 7.65 \times 10^{-4}}{(0.01)^2}$$

$$= 45.90 \text{ MN/m}^2$$

Adding,

$$\sigma = 3.75 \pm 45.90 \text{ MN/m}^2$$

$$= 49.65 \text{ MN/m}^2 \text{ (inside surface)}$$

$$= -42.15 \text{ MN/m}^2 \text{ (outside surface)}$$

(b) Total hoop stress in the circumferential direction is given by,

$$\sigma = \sigma_\theta - \sigma_c \pm \sigma_b$$

where,

$$\sigma_\theta = \frac{pr}{t} = \frac{0.5 \times 0.15}{0.01} = 7.50 \text{ MN/m}^2$$

$$\sigma_c = \frac{E}{r} \delta_{c,\, P_o\, M_o}$$

$$= \frac{E}{r} \times \frac{1}{E} (149.3 \, P_o - 4\,950 \, M_o)$$

$$= \frac{1}{0.15} (149.3 \times 3.07 \times 10^{-2} - 4\,950 \times 10^{-4})$$

72   CHEMICAL EQUIPMENT DESIGN—MECHANICAL ASPECTS

$$= 5.32 \text{ MN/m}^2$$

$$\sigma_b = \frac{6}{t^2}(\mu M_o)$$

$$= \frac{6 \times 0.3 \times 7.65 \times 10^{-4}}{(0.01)^2} = 13.77 \text{ MN/m}^2$$

Adding,

$$\sigma = 7.50 - 5.32 \pm 13.71 \text{ MN/m}^2$$

$$= 15.95 \text{ MN/m}^2 \text{ (outside surface)}$$

$$= 11.59 \text{ MN/m}^2 \text{ (inside surface)}$$

Stress intensification factor K_t is obtained dividing σ by σ_θ in both the cases. Though, the discontinuity stresses are quite large compared to normal membrane stresses, the resulting stresses are much below the allowable stress values for carbon steel or low alloy steel Therefore, for this pressure rating, the shell thickness is large. In this analysis corrosion is neglected. Local stresses can go upto yield stress value of the material without causing any damage to the vessel life.

Design Example 5.3 : For the conditions given in Example 5.2, determine the total radial stress in the flat head at its juncture with the cylindrical shell portion.

Solution : The total radial stress at the juncture is affected by P_o and M_o. It is, therefore, composed of radial stress caused by P_o and bending stress by M_o.

$$\sigma = \frac{P_o}{t_h} \pm \frac{6 M_o}{t_h^2}$$

From previous example, $P_o = 3.07 \times 10^{-2}$ and $M_o = 7.65 \times 10^{-1}$ and substituting these values,

$$\sigma = \frac{3.07 \times 10^{-2}}{2 \times 10^{-2}} \pm \frac{6 \times 7.65 \times 10^{-4}}{4 \times 10^{-4}} \text{ MN/m}^2$$

$$= 1.535 \pm 11.475 \text{ MN/m}^2$$

$$= 13.01 \text{ MN/m}^2 \text{ (inside surface)}$$

$$= 9.94 \text{ MN/m}^2 \text{ (outside surface)}$$

5.3.3 Discontinuity stresses due to differential thermal expansion

Depending upon the physical and metallurgical properties of material of construction coefficient of thermal expansion varies. For example, coefficient of thermal expansion of austenitic stainless steel is about 50 per cent more than that of ordinary carbon steel. Process requirements quite often suggest to use costly stainless steel or similar material to avoid excessive corrosion, or if in certain part of the equipment the temperature is high, ordinary carbon steel will not be suitable. As a result to achieve economy bimetallic constructions are suggested. It is often felt necessary that a carbon steel process vessel be given lining with non-corrosive metals and inlet-outlet nozzles are also made of non-corrosive metals. Another similar case can be cited when heat exchange tubes are fixed with tube plate of different material. Some times two pipes of different kinds of steel are welded together, when one part is to be at high temperature zone. For all these cases differential radial expansion will occur resulting discontinuities at the junctions. The stresses due to differential dilation of two parts can be evaluated in the same manner as those at the junction of cylindrical shell with hemispherical head discussed in section 5.3.1. Here two locations of bimetallic joints will be considered : (a) at the juncture of nozzle and vessel wall, and (b) at the junction of two equal dimensions pipes of different material. The differential dilation of two parts in all the cases will be due to their different thermal expansion and is given as follows :

$$\delta = r \triangle T (\alpha_1 - \alpha_2) \qquad \ldots(5.3.24)$$

where $\triangle T$ is the temperature change, r is the radius and α is the linear coefficient of thermal expansion. As will be seen from the following analysis, the location of the bimetallic joint influences greatly the magnitude of discontinuity stress. For the case study, the attachment of a stainless steel nozzle to a heavy wall carbon steel vessel will be considered.

(a) Location of the joint at the juncture of the nozzle with the vessel wall :

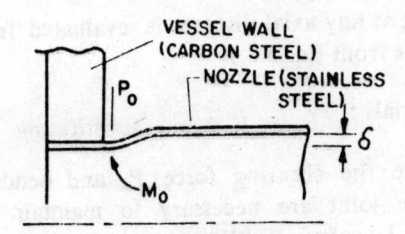

Fig. 5.4 Discontinuity at the junction of nozzle with vessel wall of dissimilar metals

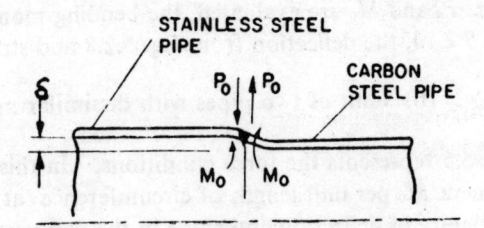

Fig. 5.5 Discontinuity at the juncture of two pipes having different materials of construction

Fig. 5.4 represents the force conditions. In this case, for obvious reason, the wall of carbon steel vessel is considered to be rigid and will not deflect or twist, so that the stainless steel nozzle must absorb the complete deflection, δ, due to differential thermal expansion between two parts and it is also prevented from rotating at the juncture of giving $\theta_{x=0} = 0$. From Eqs. 5.3.24 and 5.2.8 the deflection continuity condition gives,

$$r \Delta T (\alpha_1 - \alpha_2) = \frac{2P_o \beta}{k} D_{\beta x} - \frac{2M_o \beta^2}{k} C_{\beta x} \qquad ...(5.3.25)$$

Here, suffix 1 stands for stainless steel and suffix 2 for carbon steel. From Eq. 5.2.9 the edge slope equation gives,

$$0 = \frac{4 M_o \beta^3}{k} D_{\beta x} - \frac{2 P_o \beta^2}{k} A_{\beta x} \qquad ...(5.3.26)$$

At $x = 0$, $A_{\beta x} = C_{\beta x} = D_{\beta x} = 1$. Substituting in above equations,

$$r \Delta T (\alpha_1 - \alpha_2) = \frac{2 P_o \beta}{k} - \frac{2 M_o \beta^2}{k} \qquad ...(5.3.27)$$

and

$$0 = \frac{4 M_o \beta^3}{k} - \frac{2 P_o \beta^2}{k} \qquad ...(5.3.28)$$

Solution of above two equations gives,

$$M_o = \frac{k [r \Delta T (\alpha_1 - \alpha_2)]}{2\beta^2} \qquad ...(5.3.29)$$

$$P_o = \frac{k [r \Delta T (\alpha_1 - \alpha_2)]}{\beta} \qquad ...(5.3.30)$$

Once P_o and M_o are evaluated, the bending moment at any axial distance is evaluated from Eq. 5.2.10, the deflection from Eq. 5.2.8 and stresses from Eq. 5.3.1.

(b) Joint of two pipes with dissimilar materials :

Fig. 5.5 represents the force conditions. In this case the shearing force P_o and bending moment M_o per unit length of circumference at the joint are necessary to maintain the continuity of deflection and slope in the two parts. To satisfy the continuity requirement, the sum of the deflection of the stainless steel portion and carbon steel portion resulting from P_o and M_o given by Eq. 5.2.8 must be equal to the differential thermal expansion given by Eq. 5.3.24,

$$r \Delta T (\alpha_1 - \alpha_2) = \delta_{1, Po} - \delta_{1, Mo} + \delta_{2, Po} + \delta_{2, Mo} \qquad ...(5.3.31)$$

$$= \frac{2 P_o \beta_1}{k_1} D_{\beta x} - \frac{2 M_o \beta_1^2}{k_1} C_{\beta x} + \frac{2 P_o \beta_2}{k_2} D_{\beta x} + \frac{2 M_o \beta_2^2}{k_2} C_{\beta x} \qquad ...(5.3.32)$$

$$= 2 P_o \left[\frac{\beta_1 D_{\beta x}}{k_1} + \frac{\beta_2 D_{\beta x}}{k_2} \right] + 2 M_o \left[\frac{\beta_2^2 C_{\beta x}}{k_2} - \frac{\beta_1^2 C_{\beta x}}{k_1} \right] \qquad ...(5.3.33)$$

Now, equating the edge rotations of two parts from Eq. 5.2.9, noting $\theta_1 = \theta_2$

$$\frac{4 M_o \beta_1^3}{k_1} D_{\beta x} - \frac{2 P_o \beta_1^2}{k_1} A_{\beta x} = \frac{2 P_o \beta_2^2}{k_2} A_{\beta x} - \frac{4 M_o \beta_2^3}{k_2} D_{\beta x} \quad \ldots(5.3.34)$$

$$- 2 P_o \left[\frac{\beta_1^2 A_{\beta x}}{k_1} - \frac{\beta_2^2 A_{\beta x}}{k_2} \right] = - 4 M_o \left[\frac{\beta_1^3 D_{\beta x}}{k_1} + \frac{\beta_2^3 D_{\beta x}}{k_2} \right] \quad \ldots(5.3.35)$$

Two cases are arising here. (1) When the temperature is high, $E_1 \neq E_2$. Eqs. 5.3.33 and 5.3.35 are to be solved simultaneously for P_o and M_o. (2) When temperature is not so high, $E_1 = E_2$. Therefore, $k_1 = k_2$ and $\beta_1 = \beta_2$. This makes left hand side of Eq. 5.3.35 equal to zero, indicating $M_o = 0$ and the condition of slope continuity is provided by the action of the forces P_o only. It may be said, therefore, that the slope and deflection induced at the edges of the two parts by the forces P_o are equal. Hence the conditions of continuity are met by $M_o = 0$, and Eq. 5.3.33 gives

$$\frac{r \Delta T (\alpha_1 - \alpha_2)}{2} = \frac{2 P_o \beta}{k} D_{\beta x} \quad \ldots (5.3\;36)$$

An agreement is reached with Eq. 5.3.9, when it is noted that the left side of this equation is $\delta/2$. The stresses can then be computed in the similar fashion according to Eq. 5.3.1.

Design Example 5.4 : In order to prevent contamination of the coolant fluid by rust particles in nuclear reactor vessels fabricated of carbon steel, the inside of the vessel is frequently clad with a very thin layer of stainless steel (which does not affect its strength or deflection), and solid wall stainless steel nozzles are welded into the vessel wall because of the difficulty or inability to clad relatively small diameter nozzles internally. If a stainless steel nozzle ($\alpha_1 = 16 \times 10^{-6}$ m/m °C) of $r = 0.5$ m, $t = 0.012$ m, $\mu = 0.3$ and $E = 1.8 \times 10^5$ MN/m^2 is welded into a thick carbon steel vessel ($\alpha_2 = 12.6 \times 10^{-6}$ m/m °C) so as simulate a built in edge, determine : (a) the value of the nozzle characteristic β and (b) the flexural rigidity D. What is : (c) the shearing force and bending moment per m length of circumference at the built-in edge, and (d) what is the maximum thermal stresses at this location when the reactor vessel is operating to give a uniform increase in metal temperature of 150 °C ? The nozzle is free to expand in its axial direction.

Solution : (a) From Eq. 5.2.16

$$\beta = \frac{1.285}{\sqrt{rt}} = \frac{1.285}{\sqrt{0.5 \times 0.012}} = 16.6 \text{ m}^{-1}$$

(b) From Eq. 5.2.14

$$D = \frac{E t^3}{12 (1 - \mu^2)} = \frac{1.8 \times 10^5 \times (0.012)^3}{12 \times 0.91} = 0.028\;5 \text{ MN m}$$

(c) From Eq. 5.3.24

$$\delta = r \Delta T (a_1 - a_2) = 0.5 \times 150 \times 10^{-6}(16.0 - 12.6)$$

$$= 2.55 \times 10^{-4} \text{ m}$$

From Eq. 5.2.12

$$k = \frac{Et}{r^2} = \frac{1.8 \times 10^5 \times 0.012}{(0.5)^2} = 8\,640 \text{ MN/m}^3$$

From Eq. 5.3.29

$$M_o = \frac{k\delta}{2\beta^2} = \frac{8\,640 \times 2.55 \times 10^{-4}}{2 \times (16.6)^2}$$

$$= 4.0 \times 10^{-3} \text{ MN m/m}$$

From Eq. 5.3.30

$$P_o = \frac{k\delta}{\beta} = \frac{8\,640 \times 2.55 \times 10^{-4}}{16.6} = 0.133 \text{ MN/m}$$

(d) Longitudinal bending stress

$$\sigma_b = \pm \frac{6M_o}{t^2} = \pm \frac{6 \times 4.0 \times 10^{-3}}{(0.012)^2}$$

$$= \pm 166.7 \text{ MN/m}^2$$

Circumferential stresses

$$\sigma = -\frac{E}{r}\delta \pm \frac{6}{t^2}(\mu M_o)$$

$$= -\frac{1.8 \times 10^5 \times 2.55 \times 10^{-4}}{0.5} \pm \frac{6 \times 0.3 \times 4.0 \times 10^{-3}}{(0.012)^2}$$

$$= -91.8 \pm 50.0$$

$$= -141.8 \text{ MN/m}^2$$

Austenitic stainless steel of IS : 1570-1961 has the yield strength of around 190 MN/m² for the temperature not exceeding 150 °C. But in the problem ΔT is given 150 °C. In that case actual metal wall temperature would be around 175 °C. For this temperature yield strength is given about 160 MN/m². As the induced stresses in one case exceed this value, care should be taken against its fatigue failure if temperature fluctuates very often. For steady condition, there is no danger of failure.

Design Example 5.5: As a part of an external support system for a cylindrical nuclear reactor vessel, a narrow ring of cross-sectional area A is fastened snugly around the outside of the vessel at a distance well removed from the ends. Assuming zero clearance between the outside diameter of the vessel and the inside diameter of the ring, establish an expression for (a) the load P_o per m of circumference of the vessel, (b) the maximum bending moment M_o in the vessel wall, and (c) the maximum bending stress produced in the vessel wall, due to the radial dilation of the vessel of radius r and thickness t resulting from an internal pressure p, all material being steel, $\mu = 0.3$.

Solution: This problem is to be solved simulating the reactor vessel as infinitely long beam concentrically loaded and supported on elastic foundation.

(a) From Eq. 2.11.3 the total outward dilation of the vessel due to internal pressure when unrestrained is obtained as follows:

$$\delta = \frac{p\,r^2}{2\,t\,E}(2-\mu) \qquad \ldots(I)$$

At the ring location this total amount must be absorbed by a local decrease in the radius of the vessel, δ_v, and an increase in the radius of the ring, δ_r.

Therefore

$$\delta = \delta_v + \delta_r \qquad \ldots(II)$$

From Eqs. 5.2.4 and 5.2.12 at $x = 0$

$$\delta_v = \frac{r^2 P_o \beta}{2\,t\,E} \qquad \ldots(III)$$

Considering ring, P_o is the force per unit length of circumference and r is the radius. Referring to Fig. 3.1, the force acting on an element of the ring is $P_o r d\theta$, where $r d\theta$ is the circumferential element. Force balance on the semicircular ring gives

$$2F = 2\int_0^{\pi/2} P_o\, r \sin\phi\, d\phi = 2\, P_o\, r$$

or $F = P_o\, r$ $\qquad \ldots(IV)$

The unit stress in the ring can be obtained by dividing the force F by the cross-sectional area A of the ring and this gives,

$$\sigma_\theta = \frac{P_o\, r}{A} \qquad \ldots(V)$$

From the above expression and from Eq. 2.11.1

$$\delta_r = e_\theta\, r = \frac{P_o\, r^2}{E\,A} \qquad \ldots(VI)$$

From I, II, III and VI,

$$\frac{p r^2 (2-\mu)}{2 t E} = \frac{r^2}{2 t E} \frac{P_o \beta}{1} + \frac{P_o r^2}{E A}$$

or $\quad P_o \left(\dfrac{\beta}{2t} + \dfrac{1}{A} \right) = \dfrac{1.7 p}{2 t}$

or $\quad P_o = \dfrac{1.7 p A}{A\beta + 2t}$

but $\quad \beta = \dfrac{1.285}{\sqrt{rt}}$. Substituting.

$$P_o = \frac{0.85 \, p \, A \, \sqrt{rt}}{0.642\,5\, A + \sqrt{rt^3}}$$

(b) The maximum bending moment M_o in the vessel wall is obtained from Eq. 5.2.6a.

$$M_o = \frac{P_o}{4\beta} = 0.194\,5 \sqrt{rt} \; P_o$$

Substituting the value for P_o

$$M_o = \frac{0.165\,3 \, P \, A \, rt}{0.642\,5\, A + \sqrt{rt^3}}$$

(c) Maximum bending stress occurs in the longitudinal direction. (In the circumferential direction, $M_o = \mu \, M_x$ given in Eq. 5.2.13). Therefore,

$$\sigma_b \,(\text{max}) = \frac{6}{t^2} \, M_o$$

Substituting for M_o

$$\sigma_b \,(\text{max}) = \frac{1.167 \, p_o \sqrt{r}}{\sqrt{t^3}}$$

Substituting again for P_o

$$\sigma_b \,(\text{max}) = \frac{0.921\,9 \, A \, pr}{0.642\,5\, A t + t^2 \sqrt{rt}}$$

5.3.4 Stress concentration in the shell or nozzle due to flange moment

This is another case of discontinuity stresses. Discontinuity occurs at the junction of shell or nozzle and flange. When the flanges, are tightened through bolt action, a shear load

P_o and bending moment M_o come into existence to preserve the continuity at the juncture. As a result, the stresses are induced in the shell or nozzle wall. The magnitude of these depends upon the flange moments besides internal or external pressure. In flange design, the effect of flange moments in shell or nozzle is often disregarded. Thus for a 5 mm thick shell, a 50 mm thick ring flange is suggested only from flange stresses consideration. Design procedure for flanges will be discussed later in a separate chapter. An important contribution of the discontinuity stresses in the shell or nozzle due to flange moments at the juncture would be to decide the suitability of a particular type of flange. Simulating the flanged nozzle or shell with the semi-infinite beam on elastic foundation, it would be possible to determine the hub dimensions theoretically—from the consideration of stress concentration and the length affected by the bending moment and shear load.

CHAPTER 6

COMPENSATION FOR OPENINGS IN PROCESS EQUIPMENT

6.1 INTRODUCTION

Any process vessel must be provided with multiple openings of various dimensions at different parts. These are necessary for giving inlet and outlet connections, for providing sight glasses, manholes, drainage, for inserting shaft of the stirrer, etc. While the openings are essential for operating the vessels, these weaken the vessel parts due to development of discontinuities. Experimental evidence shows that the stress concentration at the edge of the openings in pressure vessels becomes as high as 500 per cent that of the unpierced shell under the similar operating conditions.[1] It is also observed that this high value of stress at the edge of the opening decreases sharply with the distance from it. This makes it necessary to ascertain adequate reinforcement in appropriate region. Again, excess use of reinforcing material may reduce flexibility and thus the dilation characteristic of that region may be changed with respect to the rest of the vessel. This will again induce discontinuity stresses. Following observations can be made from the results published in the literature[2] in connection with the reinforcing of the openings.

1. There is considerably high stress concentration at the edge of the openings and around them. The magnitude of the resultant stress value is much higher than the allowable stress limit for unpierced shells and sometimes the difference is as large as 500 per cent.

2. Stress concentration is maximum at the edge of the openings and diminishes to a negligibly small value beyond the area covered by twice the hole diameter.

3. It is necessary to adequately compensate all the openings above a minimum diameter to avoid the failure of the vessels in those regions.

4. Compensation can be provided by increasing the entire shell thickness so that the maximum anticipated stress at the edge of the opening is less than, or equal to, the allowable stress of the shell material. This may be the best method, but is not economical. If a sufficiently thick-walled nozzle is welded to the opening, the magnitude of the stress gets reduced to the safety limit. This type of compensation is quite efficient and also economical. Other commonly used methods are welded ring plate, combination of ring plate and nozzle, etc.

---

[1] S. B. Kantorowitsch, "Die Festigkeit der Apparate und Maschinen fuer die chemische Industrie" VEB, Berlin.

[a] See list of references at the end of this chapter.

5. Compensation should provide both rigidity and strength. It should be concentrated near the opening.

6. Hundred per cent compensation for the opening is not practicable and it is also not necessary, from safety point of view, due to redistribution of stresses which cause ultimate failure. On the other hand, there is a danger from over-reinforcing the openings with heavy ring plates causing stress concentration at the outer edge of the ring due to a stiffening effect.

6.2 TYPES OF COMPENSATION

Figure 6.1 shows the commonly used types of compensation for openings in process vessels. A short description of each of them are given below.

(a) Flued-in type elliptical man-ways are quite common in pressure vessels, such as, compressed air storage tanks, steam boilers, etc. Though this type provides efficient compensation, fabrication is difficult and costly.

(b) Flared-out or drawn openings further connected with nozzles may be preferred, because of large transition radii to avoid the stress raising effect of sharp corners ; but this is not essential, and may prove a dangerous practice if too large a radius is selected. In that case opening diameter will be much larger than actual nozzle diameter. This will enhance the amount of compensation requirement. One great advantage is that radiographic tests are possible in nozzle connections. At the vessel's juncture with the nozzle radiographying is difficult.

(c) Ring plates or pads welded to the shell and nozzle, are often used as a cheap method of compensation. The outer diameter of the pad is usually $1\frac{1}{2}-2$ times the opening diameter. High thermal stresses are often induced due to the poor heat transmission coefficient between the shell plate and the pad as both parts cannot be fused together. This effect will be further enhanced if dissimilar metals, having different coefficients of thermal expansion are used. In addition, the stress raising effect of the weld fillets may initiate cracking under cycle loading.

(d) Rim or nozzle type is likely to be the most efficient. Here the compensation is most effectively arranged at the edge of the opening, where stress concentration is maximum.

(e) Sweep-type nozzles and a combination of ring plates and nozzles as compensation are also used in practice.

It may be noted that ring plate type compensation causes a reduction in life by 30 per cent, with respect to the fatigue life of the unpierced vessel, while flush nozzles cause a reduction of only 15 per cent and protruding nozzles about 10 per cent.[3]

---

[3] M. B. Bickell and C. Ruiz, "Pressure Vessel Design and Analysis", Macmillan, London, 1967.

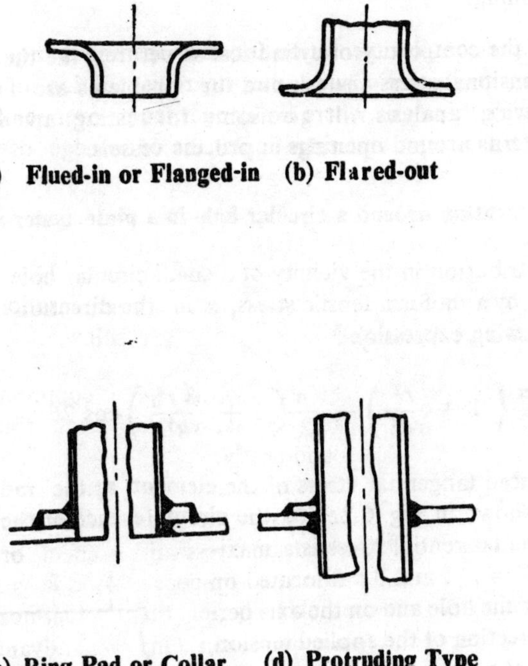

COMPENSATION FOR OPENINGS IN PROCESS EQUIPMENT

(a) Flued-in or Flanged-in   (b) Flared-out

(c) Ring Pad or Collar   (d) Protruding Type

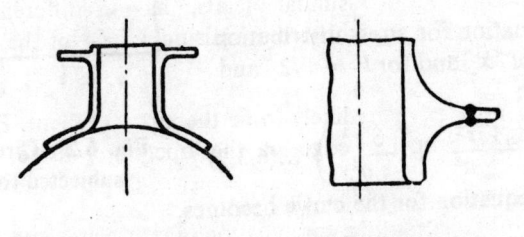

(e) Sweet Type   (f) Special Type Suitable for Radiographying

Fig. 6.1 Commonly used types of compensation for openings

6.3 THEORETICAL DETERMINATION OF STRESS PATTERNS AROUND OPENINGS

Because of the complexity of cylindrical structures, the theoretical investigations are limited to two dimensional stress analysis and the problems are mostly over-simplified.[3,4,5] However, the following analysis will give some interesting and useful informations about expected stress patterns around openings in process vessels.

6.3.1 Stress concentration around a circular hole in a plate under uniform tension

The stress distribution in the vicinity of a small circular hole of radius, r, in a plate is stretched elastically by a uniform tensile stress, σ, in the direction of the polar axis $\theta = 0$, is given by the following expression.[5]

$$\sigma_t = \frac{\sigma}{2}\left(1 + \frac{r^2}{a^2}\right) - \frac{\sigma}{2}\left(1 + \frac{3r^4}{a^4}\right)\cos 2\theta \qquad \ldots (6.3.1)$$

where σ_t is the induced tangential stress of the element at the radial distance 'a' from the centre of the hole, shown in Fig. 6.2. At the circumference of the hole, $a = r$ and $\sigma_t = \sigma(1 - 2\cos 2\theta)$. The tangential stress is a maximum at the points $\theta = \pi/2$ and $3\pi/2$ located on the circumference of the hole and on the axis perpendicular to the direction of the applied tension, σ. At these points the stress $\sigma_t = 3\sigma$. Again, for $a = r$ and $\theta = 0$ or $180°$, $\sigma_t = -\sigma$. Thus it is seen that a small hole in a plate subjected to tension in a given direction causes an increase in the stress in vicinity of the hole to a maximum value of three times that in a normal undisturbed portion of the plate.

The general equation for stress distribution curve for any value of 'a' and for $\theta = \pi/2$ and $3\pi/2$ is given below:

$$\sigma_t = K_t \sigma = \frac{\sigma}{2}\left(2 + \frac{r^2}{a^2} + \frac{3r^4}{a^4}\right) \ldots (6.3.2)$$

For $\theta = 0$ or π, the equation for the curve becomes,

$$\sigma_t = K_t \sigma = \frac{\sigma}{2}\left(\frac{r^2}{a^2} - \frac{3r^4}{a^4}\right) \qquad \ldots (6.3.3)$$

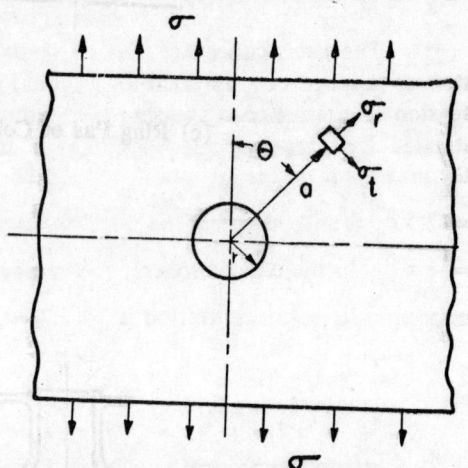

Fig. 6.2 Circular hole in a plate subjected to uniform tension

---

[4] E. O. Waters, "Reinforcement of openings in pressure vessels" Welding Journal Research Supplement, 1958.

[5] "Pressure Vessel and Piping Design" Collected papers, 1927-1959. ASME, New York.

Variation in stress in a plate containing a circular hole and subjected to uniform tension is shown in Fig. 6.3.

The exact theory of Eq. 6.3.1 is based on a small hole in an infinite plate. But from Fig. 6.3 it is seen that the effect of small hole is extremely limited and damps out rapidly. Hence, for practical purposes the formula can be used for plates of widths more than five times the hole diameter.

Furthermore, the variation in the stress concentration along a diametrical section mm parallel to the direction of tension is minimum (Fig. 6.3). At the edge of the hole it produces a compressive stress, tangent to the hole, equal to the tensile stress σ applied at the ends of plate, and damps out very rapidly with distance from the edge of the hole.

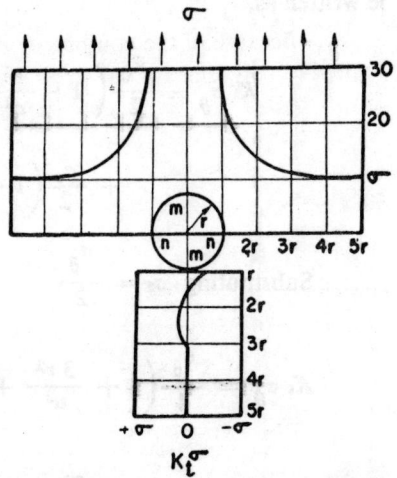

Fig. 6.3 Variation in stress around a hole in a plate subjected to tension

The stress concentrations around a circular hole in a cylinder or sphere with stresses applied by internal or external pressure can be obtained from the cases of simple tension or compression by using the method of superposition. In the case of a cylinder stressed by pressure, the longitudinal stress is half the hoop stress (Fig. 6.4), therefore the maximum stress at point n on the longitudinal axis is $3\sigma_\theta - \sigma_z = 3\sigma_\theta - \tfrac{1}{2}\sigma_\theta = 2.5\sigma_\theta$, and at point m on the circumferential axis is a $3\sigma_z - \sigma_\theta = \tfrac{3}{2}\sigma_\theta - \sigma_\theta = \tfrac{1}{2}\sigma_\theta$. In the case of sphere, the two principal stresses are equal, $\sigma_z = \sigma_\theta = \sigma$, and the maximum stress concentration is $3\sigma_\theta - \sigma_z = 3\sigma - \sigma = 2\sigma$.

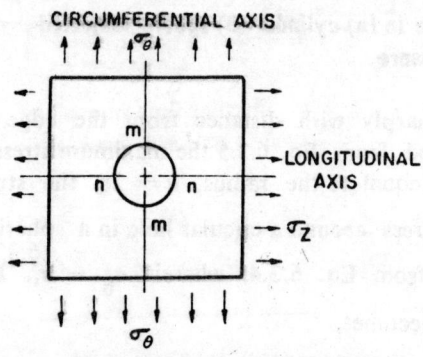

Fig. 6.4 Hole in a cylinder subjected to bi-axial stress

6.3.2 Determination of reinforcement boundaries for circular openings in cylindrical and spherical vessels

The boundaries for the addition of effective reinforcing material can be obtained by examining the stress gradient with the distance from the edge of the hole along the longitudinal axis. Fig. 6.5 shows the stress gradient from the edge of the hole for cylinder and sphere subjected to internal pressure. For a cylindrical vessel subject to internal pressure, wherein the longitudinal stress (σ_z) is one half the hoop stress (σ_θ) Eq. 6.3.1 can be written as,

$$K_t \sigma_\theta = \frac{\sigma_\theta}{2}\left(1 + \frac{r^2}{a^2}\right) - \frac{\sigma_\theta}{2}\left(1 + \frac{3r^4}{a^4}\right)\cos 2\theta \ \left(\theta = \frac{\pi}{2}\right)$$
$$+ \frac{\sigma_z}{2}\left(1 + \frac{r^2}{a^2}\right) - \frac{\sigma_z}{2}\left(1 + \frac{3r^4}{a^4}\right)\cos 2\theta \ (\theta = 0) \quad \ldots (6.3.4)$$

Substituting $\sigma_z = \dfrac{\sigma_\theta}{2}$

$$K_t \sigma_\theta = \frac{\sigma_\theta}{4}\left(4 + \frac{3r^2}{a^2} + \frac{3r^4}{a^4}\right) \quad \ldots (6.3.5)$$

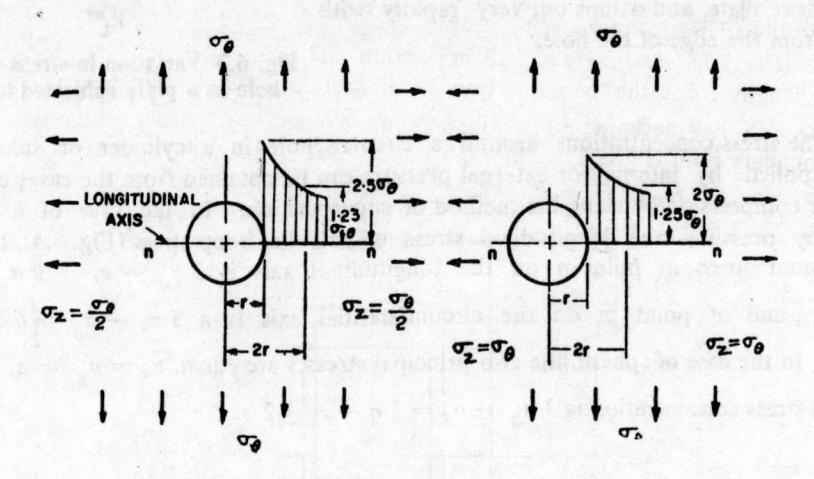

Fig. 6.5 Stress gradient around circular hole in (a) cylinder (b) sphere subjected to internal pressure

Fig. 6.5 shows that the stress decreases sharply with distance from the edge of the hole. At the edge of the hole $a = r$, and from Eq. 6.3.5 the maximum stress is 2.5 σ_θ. At a distance from the edge of the hole equal to the radius, $a = 2r$, the stress becomes 1.23 σ_θ. Similarly, the variation in stress around a circular hole in a spherical vessel subjected to internal pressure is obtained from Eq. 6.3.4, wherein $\sigma_\theta = \sigma_z$. For maximum stress distribution Eq. 6.3.4 for sphere becomes,

$$K_t \sigma_\theta = \sigma_\theta \left(1 + \frac{r^2}{a^2}\right) \quad \ldots (6.3.6)$$

A corresponding radial decrease of the stresses from the edge of the hole occurs as shown in Fig. 6.5. The stress reaches maximum at the edge of the hole equal to $2\sigma_\theta$ and falling to a value of $1.25\ \sigma_\theta$ at a distance $a = 2r$.

From the above analysis, it is observed that at a distance from the hole edge equal to the radius, the stress concentration is negligible. This is proved by experimental evidences.[6,7] Therefore, one boundary limit for the effective reinforcement area is suggested to be equal to $2d$, where d is the opening diameter (Fig. 6.6). This is accepted by various standard codes including IS.[8,9] The second boundary limit which is in the direction of perpendicular to the plate surface can be approximated from the deflection characteristics of the nozzle or ring (see Ch. 5), which is doing the reinforcement. From section 5.2.5, it can be noticed that for a cylindrical nozzle, at a distance π/β from the point of application of load, the deflection effect is practically nil. To achieve effective reinforcement from the nozzle or ring pad, the length H, which is within π/β from the surface of the nozzle, is only to be taken into account. For cylindrical nozzle $\beta = 1.285/\sqrt{\dfrac{d}{2} t_c}$, where, t_c is the corroded wall-thickness.

From this $H = 1.7\sqrt{d t_c}$. However, in section 5.2.5 it is also stated that the most effective length from the consideration of deflection characteristics would be $1/\beta$ or $0.51\sqrt{d t_c}$. IS has accepted approximately an average of these two magnitudes and the second boundary limit is given by (Fig. 6.6)

$$H_1 = H_2 = \sqrt{d t_c} \qquad \ldots (6.3.7)$$

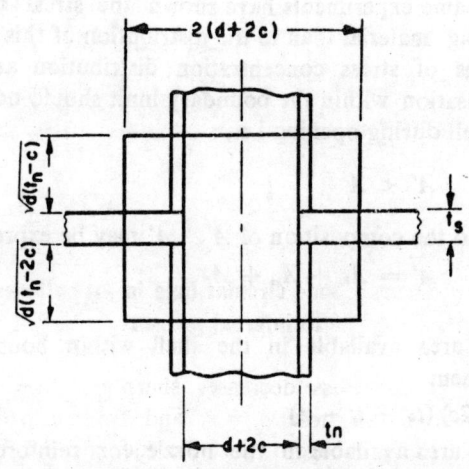

Fig. 6.6 Boundary limits of the reinforcement for circular openings in cylindrical and spherical vessels

---

[6] J. H. Taylor and E. O. Waters, "The effect of openings in pressure vessels", ASME Transactions, 1934.

[7] "Symposium on Pressure Vessel Research Towards Better Design", I.Mech.E., 1962.

[8] IS : 2825—1969, Indian Standards Institution, New Delhi.

[9] B. C. Bhattacharyya, "Requirement of compensation for openings in pressure vessel design", Chemical Age of India, 1972.

6.3.3 Determination of area to be compensated and area available for reinforcement

When an opening is made in a vessel wall, the wall becomes weaker because of the decrease in cross-sectional area perpendicular to the hoop stress direction. It is seen in Ch. 3, in designing pressure vessel, the safety against induced hoop stress is mainly considered. Therefore, in determining the area to be compensated, one is to find what basic area perpendicular to the hoop stress direction has got removed due to the opening. This is nothing but corroded opening or nozzle diameter multiplied by the theoretically calculated minimum wall-thickness from Eq. 3.3.19. One is to note that, the value of J in that equation will depend upon the position of the nozzle. If the nozzle opening is made away from any longitudinal welded seam, $J = 1$; and in case the hole is to be made on the longitudinal welded joint $J < 1$. In designing and planning process vessels, it is always preferred to avoid any opening through welded joint, being weaker section. The corroded diameter is the internal diameter of the opening plus twice the corrosion allowance. This is done because, during design worst condition is to be considered. Therefore, if 'A' is the basic area being reduced due to opening, then,

$$A = (d + 2c) t_r \qquad \ldots (6.3.8)$$

Now comes the difficult problem of determining the correct amount of reinforcement. If A' is the area of reinforcing elements, then photoelastic investigations[5] show that for A'/A varying 65 to 115 per cent, the improvement in maximum stress conditions is negligible.

Further, the same experiments have shown the stress to be less sensitive to the amount of reinforcing material than to the distribution of this material. This agrees with the theoretical findings of stress concentration distribution around openings. However, the area of compensation within the boundary limit should not be less than the basic area removed from the shell during opening i.e.,

$$A' \not< A$$

Next is to find the composition of A'. A' may be expressed as,

$$A' = A_s + A_n + A_r \qquad \ldots (6.3.9)$$

In this equation,

A_s = excess area available in the shell within boundary limit acting as reinforcement

$$= (d + 2c)(t_s - t_r - c) \qquad \ldots (6.3.10)$$

A_n = excess area available in the nozzle for reinforcement. This is when the actual nozzle wall thickness is more than the minimum thickness required from hoop stress consideration.

$$= A_o + A_i$$

A_o = area of the nozzle external to the vessel available for compensation (Fig. 6.7)

$$= 2 H_1 (t_n - t_r' - c) \qquad \ldots (6.3.11)$$

COMPENSATION FOR OPENINGS IN PROCESS EQUIPMENT

A_i = area of the nozzle inside the vessel (i.e. protruding nozzle) available for compensation (Fig. 6.7)
$$= 2 H_2 (t_n - 2c) \qquad \ldots (6.3.12)$$

For inside protrusion t_r' is zero, as there is no pressure difference; but corrosion is from both the sides. A few words about H_1 and H_2 will make the matter more clear. H_1 and H_2 are the outside and inside protrusion respectively indicating the one boundary limit.

If nozzle length outside the vessel is larger than H_1, the boundary limit, then,
$$H_1 = \sqrt{(d + 2c)(t_n - c)} \qquad \ldots (6.3.13)$$

If, on the other hand, the nozzle length outside the vessel surface is less than or equal to the height of the boundary limit, then,

H_1 = actual length of the nozzle

Similarly, if the inside protrusion of the nozzle goes beyond the boundary zone, then,
$$H_2 = \sqrt{(d + 2c)(t_n - 2c)} \qquad \ldots (6.3.14)$$

On the other hand, if inside protrusion is less or equal, then,

H_2 = actual length of protruded portion.

After calculating A_s and A_n, if it is found that $A_s + A_n \geqslant A$, then no other external reinforcement is necessary.

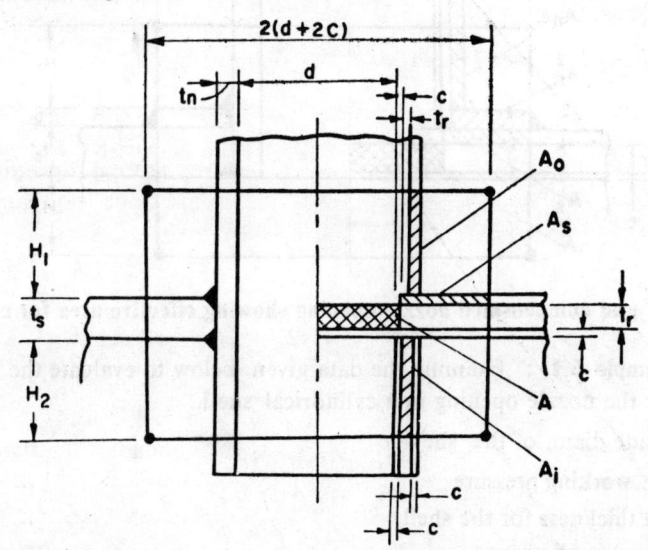

Fig. 6.7 Protruded nozzle connection showing effective area for compensation

On the other hand, if $A_s + A_n < A$, the difference in area $A - (A_s + A_n)$ is to be provided with ring pad and weldments (Fig. 6.8). Therefore

A_r = area available from ring pad and weldments within boundary limit
$$\geqslant A - (A_s + A_n)$$

CHEMICAL EQUIPMENT DESIGN—MECHANICAL ASPECTS

It is to be noted that for using as reinforcement the ring pad should be rigid enough. For that purpose, ring pad thickness should not be less than 5 mm. Further, the ring pad dimensions should be so chosen that, the area is concentrated near the edge of the opening. This gives better effectiveness to the reinforcement.

Also to be noted that if material of construction for nozzle and ring pad is having different allowable stress values than for shell, the following corrections for A_n and A_r to be made. In that case Eq. 6.3.9 is to be modified as follows,

$$A' = A_s + A_n \frac{f_n}{f_s} + A_r \frac{f_r}{f_s} \qquad \ldots (6.3.9a)$$

where,

f_s = allowable stress for shell material
f_n = allowable stress for nozzle material
f_r = allowable stress for ring pad material

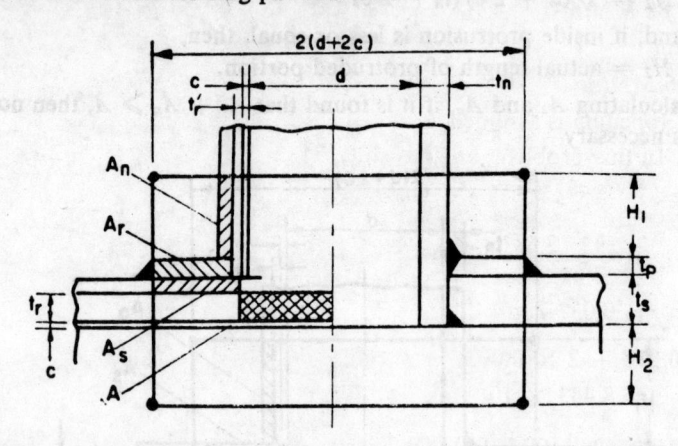

Fig. 6.8 Ring pad compensated nozzle opening showing effective area for compensation

Design Example 6.1 : Examine the data given below to evaluate the requirement of compensation for the nozzle opening in a cylindrical shell.

Outside diam. of the shell	2 m
Max. working pressure	3.5 MN/m²
Wall thickness for the shell	0.05 m
Corrosion allowance	3×10^{-3} m
Weld joint efficiency factor (Class I)	1
Allowable stress (IS : 2002-1962-2A)	96 MN/m²
Outside diam. of nozzle (seamless)	0 25 m
Nozzle-wall thickness	0.016 m
Inside protrution of nozzle—not desired	
Length of nozzle above surface	0.1 m

COMPENSATION FOR OPENINGS IN PROCESS EQUIPMENT

Solution : From Eq. 6.3.8 area to be compensated is found below
$$A = (d + 2c)t_r$$

From the given data,
$$d = (0.25 - 2 \times 0.016) \text{ m}$$
$$= 0.218 \text{ m}$$
$$c = 3 \times 10^{-3} \text{ m}$$

From Eq. 3.3.19
$$t_r = \frac{p D_o}{2fJ + p}$$

where,

p = design pressure
$= 3.5 \times 1.05 \text{ MN/m}^2$
$= 3.68 \text{ MN/m}^2$
$D_o = 2 \text{ m}$
$f = 96 \text{ MN/m}^2$
$J = 1$ (opening is assumed away from welded joint.
In this problem, however, there is no difference)

Substituting,
$$t_r = \frac{3.68 \times 2}{2 \times 96 \times 1 + 3.68} \text{ m}$$
$$= 0.037\,7 \text{ m}$$

Hence, $A = (0.218 + 2 \times 0.003) \times 0.037\,7 \text{ m}^2$
$$= 8.443 \times 10^{-3} \text{ m}^2$$

Area available from shell for reinforcement
$$A_s = (d + 2c)(t_s - t_r - c)$$
$$= (0.218 + 2 \times 0.003)(0.05 - 0.037\,7 - 0.003)$$
$$= 2.083 \times 10^{-3} \text{ m}^2$$

Area available from nozzle for reinforcement
$$A_n = A_o \text{ (no inside protrution)}$$
$$= 2 H_1 (t_n - t_r' - c)$$

Where, $t_n = 0.016 \text{ m}$
$$t_r' = \frac{p d_o}{2fJ + p}$$
$$= \frac{3.68 \times 0.25}{2 \times 96 \times 1 + 3.68}$$
$$= 0.004\,7 \text{ m}$$

From Eq. 6.3.13

$$H_1 = \sqrt{(d + 2c)(t_n - c)}$$
$$= \sqrt{(0.218 + 2 \times 0.003)(0.016 - 0.003)}$$
$$= 0.054 \text{ m}$$

Actual length of nozzle above shell surface is larger than 0.054 m
Hence,

$$A_n = 2 \times 0.054 (0.016 - 0.004\ 7 - 0.003)$$
$$= 8.964 \times 10^{-4} \text{ m}^2$$

Reinforcement area available from shell and nozzle is

$$A_s + A_n = (2.083 \times 10^{-3} + 0.896\ 4 \times 10^{-3}) \text{ m}^2$$
$$= 2.979\ 4 \times 10^{-3} \text{ m}^2$$

Area remained to be compensated is

$$A - (A_s + A_n) = (8.443 - 2.979\ 4) \times 10^{-3} \text{ m}^2$$
$$= 5.463\ 6 \times 10^{-3} \text{ m}^2$$

This difference is to be made up by ring pad, i.e.,

$$A_r \geq 5.463\ 6 \times 10^{-3} \text{ m}^2$$
$$= \{2(d + 2c) - (d + 2c + 2 t_r')\} t_p$$
$$= \{2(0.218 + 2 \times 0.003) - 0.233\} t_p$$

Hence,

$$t_p = \text{Thickness of ring pad}$$
$$\geq 0.025 \text{ m}$$

It is to be noted that if ring pad is used, H_1 is measured from the surface of the ring pad (Fig. 6.8). If actual length of nozzle is more than $H_1 + t_p$, full value of H_1 can be used for calculating A_n. Otherwise, H_1 should be substituted by actual nozzle length minus t_p. In this problem, actual length of nozzle = 0.1 m and $H_1 + t_p = (0.054 + 0.025)$ m = 0.079 m. Therefore, $H_1 = 0.054$ is used for calculating A_n.

Ring pad dimensions:

Inner diameter = d_o = 0.25 m
Outer diameter = $2(d + 2c)$ = 0.448 m
Thickness = 0.025 m

As the plate thickness is sufficiently large (0.025 m) to be rigid, it is not necessary to increase it further by reducing the outer diameter of the rigid pad. In this case, maximum effective outer diameter for the ring pad is used. If thickness of the pad would be less, it would be preferable to reduce the outer diameter to increase the thickness, so that the pad becomes rigid.

6.4 UNCOMPENSATED OPENINGS

It is not always necessary that any opening is to be externally reinforced. This is because, the wall thickness of the vessels finally decided is larger than the theoretically required. This extra thickness provides strength to the vessel walls to withstand stress concentration due to opening to some extent. The magnitude is, therefore, depends upon the diameter of the vessel D_o, thickness of the shell t, pressure p and allowable stress value of the material. All these factors will influence the size of the opening diameter that can be left uncompensated.

If t_s is the actual shell thickness, then $p D_o/2 (t_s - c)$ is the magnitude of the stress induced in the unpierced shell under internal pressure p. This is obviously less than the allowable stress value of the shell by a factor $t_r/(t_s - c)$. This difference in stress values between allowable and induced is the indication of the reserved strength of the shell to take care of the weakening effect caused due to openings to certain limit. It is seen from the earlier sections, if openings are made in the shell, its allowable strength is reduced considerably compared to the unpierced portion. This deficiency in strength is made by providing extra thickness as discussed in Section 6.3. By this way the enhanced stress level around the opening is brought down to the allowable stress value of the shell theoretically. For the calculation of the reinforcement area available from the shell, it is both related to the opening diameter d and shell thickness t_s by,

$$A_s = (d + 2c)(t_s - t_r - c) \quad \ldots(6.3.10)$$

Again, area to be compensated is given by,

$$A = (d + 2c) t_r \quad \ldots(6.3.8)$$

By analysing these two expressions, it is found that the ratio $t_r/(t_s - c)$ is an important factor to determine the size of opening which need not be compensated. On the other hand it can be said that smaller the value of $t_r/(t_s - c)$, larger can be the opening diameter without any external reinforcement. Let a factor K be defined as :

$$K = \frac{t_r}{(t_s - c)}$$

Then,

$$\frac{p D_o}{2(t_s - c)} = Kf \quad \ldots(6.4.1)$$

From Eq. 6.4.1, (Kf) is the magnitude of the induced stress in the unpierced shell due to internal pressure p. If any hole is made in the shell, stress concentration near the hole edge will be increased and the induced stress correlation for the portion near the opening or hole will be different. According to IS : 2825 – 1969, near the opening the Eq. 6.4.1 becomes,

$$\frac{p D_o}{1.82(t_s - c)} = K' f$$

or

$$K' = \frac{p D_o}{1.82 f (t_s - c)} \quad \ldots(6.4.2)$$

Stress enhancement around the opening is purely local in nature and this need not be nullified totally (Section 6.1) from this consideration, for,

$K' = 1$ or a little over 1, an opening diameter upto 0.05 m need not be compensated.

$K' \ll 1$, larger opening diameter upto 0.2 m can remain unreinforced depending upon the shell diameter, and the relationship is graphically presented in IS: 2825—1969.

Another rational approach for determining the uncompensated opening diameters is developed from the following considerations.[10] It is seen from Section 6.3, opening causes stress concentration in the shell. As a result near the opening yielding of the shell is expected at lower pressure than that required at the unpierced section. From this concept a weakening factor ϕ is defined as,

$$\phi = \frac{p_{0.2}}{p_y}$$

where, $p_{0.2}$ = pressure required to cause 0.2% permanent deformation near the opening.

p_y = pressure required to yield the unpierced shell.

If this weakening factor is introduced in Eq. 3.3.19, one can determine the theoretical shell thickness for uncompensated opening. This gives,

$$t = \frac{p D_o}{2 f J \phi + p} \qquad \ldots(6.4.3)$$

If opening is made away from welded joints, $J = 1$.

It is quite obvious that the weakening factor ϕ will depend on the opening diameter d_o and shell deflection characteristic $\sqrt{D_o (t_s - c)}$ where, t_s is the actual shell thickness and c is the corrosion allowance, ϕ as a function of $d_o / \sqrt{D_o (t_s - c)}$ is presented graphically in German code, for unfired pressure vessels (AD-Merkblaetter). A few values are given below:

Table 6.1

$d_o / \sqrt{D_o (t_s - c)}$	ϕ	$D_o / \sqrt{D_o (t_s - c)}$	ϕ
0.0	1.000	4.0	0.245
0.25	0.900	4.5	0.215
0.5	0.785	5.0	0.180
0.75	0.700	5.5	0.155
1.0	0.645	6.0	0.130
1.5	0.545	6.5	0.115
2.0	0.465	7.0	0.090
2.5	0.390	7.5	0.080
3.0	0.340	8.0	0.075
3.5	0.285		

[10] L. Winn, H. Montag and H. Hennecken, "Die Berechnung von Ausschnittversstaer-kungen" Technische Ueberwachung, 1 (1960), 409.

Design Example 6.2 : Examine the possibility of making one 0.05 m diameter uncompensated opening in the shell of Example 6.1.

Solution : From Eq. 6.4.2

$$K' = \frac{p D_o}{1.82 f (t_s - c)}$$

where,

$p = 3.68$ MN/m^2
$D_o = 2$ m
$f = 96$ MN/m^2
$t_s = 0.05$ m
$c = 0.003$ m

substituting,

$$K' = \frac{3.68 \times 2}{1.82 \times 96 (0.05 - 0.003)}$$
$$= 0.892\ 5$$

As K' is less than 1.0, as per IS : 2825 –1969 a 0.05 m diameter uncompensated opening can be made.

Let the question be examined as per German code.

$$\frac{d_o}{\sqrt{D_o (t_s - c)}} = \frac{0.05}{\sqrt{2 (0.05 - 0.003)}} = 0.163$$

From Table 6.1

$$\phi = 0.935$$

substituting in Eq. 6.4.3

$$t = \frac{p D_o}{2 f J \phi + p}$$
$$= \frac{3.68 \times 2}{2 \times 96 \times 1 \times 0.935 + 3.68}$$
$$= 0.040\ 2 \text{ m}$$

t is less than $(t_s - c)$ i.e. 0.047 m, which is available in the shell. Therefore, as per German code a 0.05 m uncompensated opening is possible.

It may be mentioned here that according to ASME section VIII, all openings not exceeding 0.05 m diameter need not be compensated.

6.5 DETERMINATION OF COMPENSATION REQUIREMENT FOR OPENINGS IN HEADS

Procedure is same as described in Section 6.3.3. Here the evaluation of t_r is to be made according to the following norm.

(a) For dished and hemispherical ends

If the opening and its compensation are located entirely within the spherical portion of a dished end, t_r is the thickness required for a sphere having a radius equal to the crown radius.

(b) For semi-ellipsoidal end

When the opening and its compensation are in an ellipsoidal end and are located entirely within a circle having a radius, measured from the centre of the end, of 0.40 of the shell diameter, t_r is the thickness required for a sphere having a radius R, derived from the following table :

Table 6 2

h_i/D_i	R_i/D_i	h_i/D_i	R_i/D_i
0 167	1.36	0.277	0.81
0.178	1.27	0.312	0.73
0.192	1.18	0.357	0.65
0.207	1.08	0.40	0.59
0.227	0.99	0.45	0.54
0.25	0.90	0.50	0.50

6.6 COMPENSATION FOR MULTIPLE OPENINGS

It is not uncommon in process equipment to find several openings grouped together over a comparatively small region. The problem of determining the effect of the interaction between openings on the stress distribution then arises. This interaction depends on the distance between the centre lines of each pair of openings, or pitch. The interaction will be negligible when the pitch becomes sufficiently large, in which case, each opening can be treated in isolation.

In case of multiple openings grouped nearby, it is to be examined, if they may be regarded as isolated openings. In this question the deflection characteristic of the shell β (Ch. 5) comes for consideration.

As a conservative estimation it can be said that there will be negligible interaction between two openings, if their edge distance is roughly $2\pi/\beta$. This value is given by German code[2] as :

$$L - d \geqslant 3 \sqrt{D_o (t_s - c)} \qquad \ldots (6.6.1)$$

where L is the pitch and d is the inside diameter of the large opening (Fig. 6.9).

According to Mershon,[2] the interaction between two openings is virtually negligible when,

$$L - d \geqslant 0.86 \sqrt{D_o(t_s - c)} \qquad ...(6.6.2)$$

This magnitude is little less than $2/\beta$ as suggested by Harvey (Ch. 5).

As per IS : 2825—1969 the openings spaced apart a distance not less than

$$L = \frac{d}{1 - 0.95 K} \qquad ...(6.6.3)$$

but in no case less than twice the diameter of the larger opening may be regarded as isolated openings.

At present the design of pressure vessels with multiple openings is based on the area replacement method.[3]

When the distance between two openings is smaller, the effect of interaction is to be examined. One should note that the cross-sectional area between two opening edges must be large enough to carry the induced load.

In the region indicated by the centre distance L, the effective cross-sectional area A_e, for the nozzle type reinforcement is given by (Fig. 6.9),

A_e = shell area + ½ nozzle area

$$A_e \leqslant \left(L - \frac{d_1 + d_2}{2}\right)(t_s - c) + \tfrac{1}{2}[H_1(t_n + t_n' - 2c) + H_2(t_n + t_n' - 4c)] \quad ...(6.6.4)$$

(a) If the openings are along longitudinal direction

$$\frac{p\, D_i\, L}{2\, A_e} \leqslant f \qquad ...(6.6.5)$$

(b) If openings are along circumference or on sphere

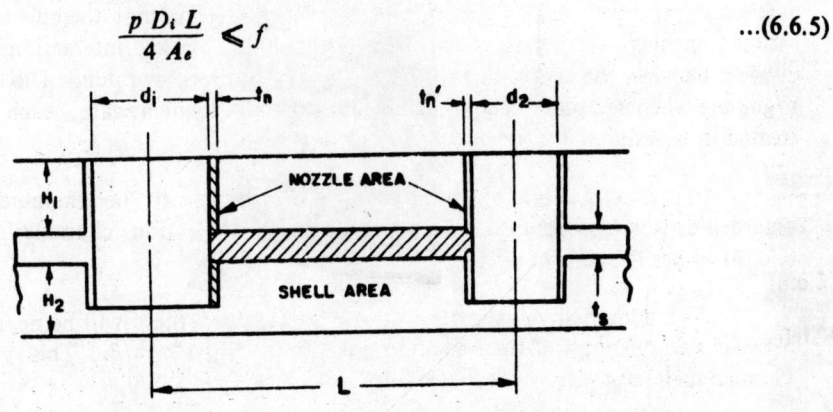

$$\frac{p\, D_i\, L}{4\, A_e} \leqslant f \qquad ...(6.6.5)$$

Fig. 6.9 Effective area for compensation of multiple openings

REFERENCES:

2.1 E. Siebel and F. Koerber, "Versuche ueber die Anstrengung und die Formaenderungen gewoelbter Kesselboden mit und ohne Manloch bei der Beanspruchung durch inneren Druck," Kaiser-Wilhelm Institut fuer Eisenforschung zu Dussel-dorf, Mitteilungen, 7 (10), 113-177 ; 8 (1), 1-51.

2.2 E. Siebel and H. Hauser, "Versuche ueber die Beanspruchung von Zylindern mit eingeschweissten Stutzen", Tech. Mitt. aus dem Dempfkessel—, Behaelter—, und Rohrleitungsbau. Duesseldorf, May, 1955.

2.3 E. Siebel and S. Schwaigerer, "Untersuchungen Ueber das Festigkeitsverhalten ausgehalster Abzweigstueck" Mitt. der Vereinigten Rohr-leitungsbau GmbH. Dusseldorf, Jan. 1954.

2.4 "Design of Pressure Vessel Nozzles", Brit. Welding, J., 9 (1962), 1500.

CHAPTER 7

DESIGN OF NON-STANDARD FLANGES

7.1 INTRODUCTION

To provide leak-proof connections between two pieces of pipes, pipes and nozzles, shells and ends or between two parts of vessels, they can be either welded, riveted or flanged as the case may be. First two methods give permanent joints. Flange joint on the other hand permits disassembly and removal or cleaning of internal parts.

Different types of flanges commonly used are :

1. Welding-neck (Fig. 7.1 a)
2. Slip-on (Fig. 7.1 b)
3. Lap-joint (Fig. 7.1 c)
4. Blind (Fig. 7.1 d)

Welding-neck flanges differ from other types in that they have a long, tapered hub between the flange ring and the weld joint with the pipe or shell. This hub provides a more gradual transition from the flange ring thickness to the pipe wall thickness, thereby decreasing the discontinuity stresses and consequently increasing the strength of the flange. This type is the strongest among the different kinds of flanges. Therefore, this type of flanges is preferred for extreme service conditions such as : repeated bending from line expansion or other forces, wide fluctuations in pressure or temperature, high pressure, high temperature, and subzero temperature. These flanges are recommended for the handling of costly, flammable, or explosive fluids, where failure or leakage of a flange joint might bring disastrous consequences. This type of flange is directly welded to the pipe or nozzle. Flanges must be properly aligned while welding to the nozzle or shell. Flanges are normally manufactured by forging.

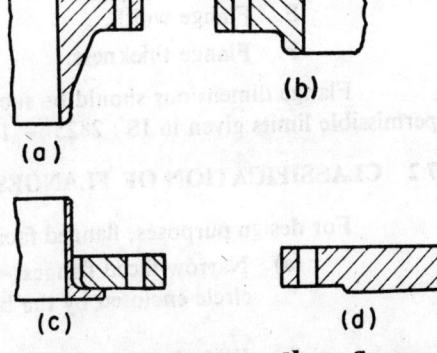

Fig. 7.1 Commonly used type flanges

The slip-on flange is simply slipped on to the pipe or shell and lap welded. This type of flange is widely used because of its greater ease of alignment in welding assembly and because of its low initial cost. The strength of this flange as calculated from internal pressure considerations is approximately 2/3 that of a corresponding welding-neck type of flange. The use of this type of flange should be limited to moderate services where pressure fluctuations, temperature fluctuations, vibrations, and shock are not expected to be severe. The fatigue life of this flange is approximately 1/3 that of a welding-neck flange.

Lap-joint flanges are usually used with lap-joint stub. The combined cost of the two parts is more than the cost of welding-neck flange of the same size and pressure rating. These flanges are as good as the slip-on flange to withstand the pressure without leaking. These flanges have the disadvantage of having only about 10% of the fatigue life of welding-neck flanges. The primary advantages are—the bolt holes are easily aligned and this simplifies the erection of vessels of large diameter and unusually stiff piping. These flanges are also useful in cases where frequent dismantling for cleaning or inspection is necessary, or where it is necessary to rotate the pipe.

Blind flanges are extensively used to blank off process vessel openings such as hand-holes and inspection ports. They are also used to block off the ends of piping and valves. Blind flanges absorb high bending stresses.

The design of flange involves :

A. Selection of the gasket (material, type and dimensions),
B. Flange facing,
C. Bolting,
D. Hub proportions,
E. Flange width,
F. Flange thickness.

Flange dimensions should be such that the stresses in the flange do not exceed the permissible limits given in IS : 2825 – 1969.

7.2 CLASSIFICATION OF FLANGES

For design purposes, flanged facings are classified either as :
(a) Narrow-faced flanges — where all the face contact area lies inside the circle enclosed by the bolt holes.
or
(b) Wide-faced flanges — with face contact area extend outside the circle enclosed by the bolt holes.

Narrow-faced flanges ensure better leak-proof joint, as gasket in between two flanges can be pressed properly. This type of flanges are, therefore, mostly used in process vessel connections. IS : 2825 – 1969 gives design procedure for this type of flange joints.

7.2.1 Types of narrow-faced flanges

Types of flange faces are decided from the point of gasket application and construction of the faces. A great variety of flange faces are in use. Fig. 7.2 shows most commonly used ones,

DESIGN OF NON-STANDARD FLANGES

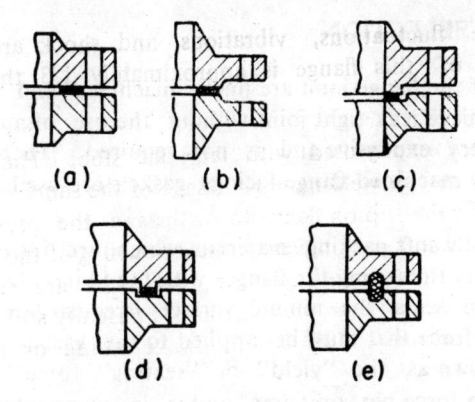

Fig. 7.2 Commonly used flange faces

(a) Plain face (Fig. 7.2 a)

Both the connecting flanges are identical. Simple in construction and inexpensive. But there is the danger of gasket blow out. Good for low pressure service.

(b) Raised face (Fig. 7.2 b)

Characteristically same as plain faced flanges. But this type ensures better compression load on gasket as there is no contact between metal faces at the point of load application, even when the gasket is totally pressed.

(c) Male and female (Fig. 7.2 c)

One face is raised and the other one is recessed. This type is more expensive than raise face. This type is suitable for high pressure operation as blow out of gasket is prevented. But at high vacuum or large external pressure, gasket may be squeezed in to the inner diameter of the vessel.

(d) Tongue and groove (Fig. 7.2 d)

This type is most suitable for high pressure as well as for high vacuum operation. There is no possibility of blowing out of squeezing in of the gasket. This type is very expensive. The tongue may be damaged while dismantling.

(e) Ring type (Fig. 7.2 e)

This type has got groove in both the faces. In this type also blow out and squeezing in of the gasket are eliminated. As construction is comparatively simpler, this type is less expensive than tongue and groove. But for this type specially made gaskets are necessary.

7.3 GASKET AND ITS SELECTION

If both the faces of the flange-joint are finely machined and free from any irregularities, it is possible to achieve leak-tight joint without the use of any external agency like gasket. But this will be very expensive and is not required. Therefore, to obtain leak-proof joint with reasonably machined flange facings, gaskets are used.

Gaskets are normally soft packing materials which are introduced in between the flange faces. While initially tightening the flanges with the bolts, the gasket gets deformed under compressive load and seals the minute surface irregularities to prevent leakage of the fluid. The amount of force that must be applied to the gasket to flow and seal the surface irregularities is known as the "yield" or "seating" force. This force is usually, expressed as a unit stress in force per unit area and is independent of the pressure inside the vessel. Thus, this yield stress represents the minimum load that must be applied to the gasket to seat it even though very low pressure are used in the vessel.

Upon the application of internal pressure in the vessel, an end force tends to separate the flanges and to decrease the unit stress on the gasket. If the hydrostatic end force is sufficiently large that the difference between it and bolt-load force reduces the gasket load below a critical value, leakage will occur. It may also be possible that the gasket is blown out due to internal pressure when contact pressure is very low. Therefore, the residual gasket force which is equal to gasket seating force minus the hydrostatic pressure force must not be less than that required to prevent leakage of the internal fluid under operating pressure.

The ratio of the gasket stress, when the vessel is under pressure, to the internal pressure is termed the "gasket factor". The gasket factor is a property of the gasket material and construction and is independent of the internal pressure over a wide range of pressures. In selecting the proper gasket for an existing closure, one of the first step should involve the determination of the total amount of force necessary to make the gasket yield and to maintain a tight seal under operating conditions.

Besides pressure, the temperature and chemical nature of the confined material and also the relative ease and economy of installation and maintenance are to be taken into consideration for the selection of gasket. The gasket material should be as serviceable as the materials of the equipment with which the gasket is used, but must not form a permanent bond with the flange faces. Some common types of gasket materials with gasket factor, m, and the minimum design seating stress, y, are listed in Table 7.1. The effective width of the gasket, b, for various types of facings is given in Table 7.2. The effective width, which is less than the actual width, N, of the gasket, depends on the facing construction of the flanges. Load effect on gasket will be different for different types of flange facings. Therefore, to determine the required joint-contact surfaces compression load necessary to ensure at tight joint at operating or working conditions, the effective gasket yield width, b, is first to be determined from Table 7.2.

DESIGN OF NON-STANDARD FLANGES

Table 7.1 Gasket Materials and suggested Values of m and y
(IS : 2825 — 1969)

Gasket Material		Gasket Factor m	Min. Design Seating Stress, y, MN/m^2	Min. Actual Gasket Width (mm)
Vulcanized rubber sheet hardness above 70 IHRD		1.00	1.38	10
Asbestos with a suitable binder for operating conditions	3.2 mm thick	2.00	11.00	10
	1.6 ,,	2.75	25.50	10
	0.8 ,,	3.50	44.85	10
Rubber with cotton fabric insertion		1.25	2.76	10
Rubber with asbestos fabric insertion, with or without wire reinforcement	3-ply	2.25	15.25	10
	2- ,,	2.50	20.00	10
	1- ,,	2.75	25.50	10
Vegetable fibre		1.75	7.56	10
Spiral-wound metal, asbestos filled	Carbon Steel	2.50	20.00	10
	S.S. or monel metal	3.00	31.00	10
Corrugated metal, asbestos inserted or Asbestos filled corrugated metal jacket	Soft Al	2.50	20.00	10
	Soft Cu or brass	2.75	25.00	10
	Iron or soft steel	3.00	31.00	10
	Monel metal	3.25	38.00	10
	S.S.	3.50	45.00	10
Corrugated metal	Soft Al	2.75	25.50	10
	Soft Cu or brass	3.00	31.00	10
	Iron or soft steel	3.25	38.00	10
	Monel metal	3.50	45.00	10
	S.S.	3.75	52.50	10
Asbestos filled flat metal jacket	Soft Al	3.25	38.00	10
	Soft Cu or brass	3.50	45.00	10
	Iron or soft steel	3.75	52.05	10
	Monel metal	3.50	55.00	10
	S.S.	3.75	62.50	10
Solid flat metal	Soft Al	4.00	61.00	6
	Soft Cu or brass	4.75	90.00	6
	Iron or soft steel	5.50	125.00	6
	Monel metal	6.00	150.00	6
	S.S.	6.50	180.00	6
Ring joint	Iron or soft steel	5.50	125.00	6
	Monel metal	6.00	150.00	6
	S.S.	6.50	180.00	6

Table 7.2 Effective Gasket Width
(IS : 2825 − 1969)

Type of flange facing	Basic gasket seating width, b_o	Effective gasket seating width, b
Plain face (Fig. 7.2 a)	$N/2$, where N is the actual gasket width in contact	$b = b_o$, when $b_o \leq 6.3$ mm
Raised face (Fig. 7.2 b)	$N/2$	$b = 2.5 \sqrt{b_o}$, when $b_o > 6.3$ mm
Male and female (Fig. 7.2 c)	$N/2$	
Tongue and groove (Fig. 7.2 d)	$(N + W)/4$, where W is the width of the tongue and N is the width of the groove.	
Ring type (i.e. groove in both the faces, Fig. 7.2 e)	$W/8$, where W is the width of the ring gasket	

7.3.1 Temperature and pressure considerations

For temperatures up to 400 °C and pressures upto 2 MN/m², temperature is the controlling factor in gasket selection. For pressures below 1 MN/m² and temperatures below 80 °C, any of the gasket materials given in Table 7.1 are applicable.

Following Table gives commonly used gasket materials at various temperature and pressure ranges.

Table 7.3 Gasket Selection[1]

Temperature °C	Pressure MN/m²	Commonly used gasket materials
Below 80	Below 1	Asbestos, fibre, or rubber sheet
Up to 250	Up to 2	Compressed asbestos sheet and various metallic reinforced asbestos sheets and cloths.
Up to 400	Up to 2	Corrugated metal-asbestos gaskets and plain iron, aluminium, copper and Monel sheet gaskets,
Exceeding 400	Exceeding 2	Plain metal gaskets

[1] H. C. Hesse and J. H. Rushton, "Process Equipment Design", East-West Press Pvt. Ltd., New Delhi, 1964.

DESIGN OF NON-STANDARD FLANGES

In the last case the pressure becomes the deciding factor in the gasket selection. But the metals and alloys used must have softening or plastic flow temperatures well above the operating temperature.

The plain face flange is used extensively for temperatures upto 250 °C and pressures upto 1 MN/m².

7.3.2 Gasket dimensions

Besides temperature, pressure and corrosive nature of the confined fluids, the selection of gasket material is also based upon the gasket width. If the gasket is made too narrow, the unit stress on it will be excessive. If, again, the gasket is made too wide, the bolt load will be unnecessarily increased. For preliminary estimate of the gasket diameters following approach can be adopted. As mentioned earlier, the residual gasket force can not be less than that required to prevent leakage of the internal fluid under operating pressure, then, (Gasket seating force) − (Hydrostatic pressure force)

= (Residual gasket force)

Let d_o and d_i are the outer and inner diameters of the gasket and y, p and m are minimum design yield stress, internal pressure and gasket factor respectively. Then,

$$\frac{\pi}{4}(d_o^2 - d_i^2)y - \frac{\pi d_o^2}{4}p = \frac{\pi}{4}(d_o^2 - d_i^2)pm \quad \ldots (7.3.1)$$

Although the above equation disregards the elastic deformation of bolts, gasket and flanges, this relationship is useful for initial proportioning of the gaskets. Eq. 7.3.1 may be oriented as follows:

$$\frac{d_o}{d_i} = \left(\frac{y - pm}{y - p(m+1)}\right)^{\frac{1}{2}} \quad \ldots (7.3.2)$$

The product pm is the unit load required to compress the gasket under operating condition. Generally a gasket seating stress larger than y should not be used, as this may lead to the crushing of the gasket. If such condition is required, this should be limited to the system like tongue and groove joints.

IS : 4870 — 1968 gives the following information about gasket thickness and width of the gasket.

Thickness (mm)	Width (mm)
3	Up to 20
4	Over 20 and up to 30
5	Over 30

Thickness smaller than 3 mm can be used, if larger gasket seating stress is desired.

7.4 SELECTION OF BOLT SPACING

Bolt spacing is the distance of the centres of two adjacent bolts. For practical reasons bolt spacing should not be too small or too large. If it is too small sufficient wrench or spanner clearance will not be available for tightening the bolts and also the ligament efficiency of the flange will be reduced which may cause crack in the flange between two bolt holes. On the other hand if the bolt spacing is too large, proper compression of the gasket all along the circumference will not occur due to deflection of flange. Maximum bolt spacing for a tight joint can be determined from the following empirical relationship.[1]

$$\text{Bolt spacing (max.)} = 2d + \frac{6t}{m + 0.5} \qquad \ldots(7.4.1)$$

Where, d = bolt diameter,
  m = gasket factor,
  t = flange thickness

The minimum bolt spacing should not be less than $2.5\,d$ for smaller bolt diameter.

Table 7.4 gives the recommended bolt spacings and also the radial distance R from the bolt circle to the extreme outer edge of the hub or nozzle or weld, the highest allowable value for the radius of curvature r_c of the rounded corner between hub and ring and the distance of the bolt circle from the flange outer edge $(A - C)/2$.

Table 7.4   Recommended Bolt Spacings[2]
(All in mm)

Bolt Diam	B_s Bolt Spacing	R (minimum)	r_c (maximum)	$(A - C)/2$
M 8 × 1	—	—	—	—
M 10 × 1	—	—	—	—
M 12 × 1.5	30—75	20	6	16
M 14 × 1.5	35—75	22	8	17
M 16 × 1.5	40—75	25	10	18
M 18 × 2	45—75	27	10	20
M 20 × 2	50—75	30	10	21
M 22 × 2	55—75	33	10	23
M 24 × 2	60—75	35	11	26
M 27 × 2	68—75	38	11	28
M 30 × 2	75	44	14	30
M 33 × 2	77	47	14	33
M 36 × 3	80	50	15	37
M 39 × 3	86	52	15	40
M 42 × 3	91	55	15	42
M 45 × 3	96	57	15	44
M 48 × 3	102	61	15	48
M 52 × 3	110	65	17	52
M 56 × 4	118	69	17	56
M 60 × 4	126	75	20	59
M 64 × 4	134	80	20	62
M 68 × 4	142	85	21	66
M 72 × 4	150	89	21	69
M 76 × 4	158	93	23	72
M 80 × 4	166	96	23	75
M 90 × 4	—	—	—	—
M 100 × 4	—	—	—	—

[2] S. B. Kantorowitsch, "Die Festigkeit der Apparate und Maschinen fuer die Chemische Industrie", VEB Verlag Technik Berlin, 1955.

It is recommended in IS : 2825 — 1969 that the bolts and studs for fastening flanges should have a nominal diameter of not less than 12 mm under normal cases. If smaller diameters are used, bolting materials should be of alloy steel to avoid overstressing of smaller diameter bolts.

To achieve uniform tightening the number of bolts should be in multiple of 4. It is also recommended that the outer diameter of the flange must be at least 20 mm larger than the sum of the bolt-circle and bolt diameter to accommodate the bolt head. Bolt hole diameter in the flange should be 2-3 mm larger than the bolt diameter.

For welded neck flanges the hub thickness may be 0.5-0.75 times the flange thickness.

In calculation minimum thickness of the flange is to be taken. If there is a groove in the flange-ring, the thickness t is to be taken from the base of the groove.

To reduce the stresses in the flange, bolt-circle diameter should be as small as practicable. The practical minimum bolt-circle diameter will be calculated either on the requirement of satisfying the radial clearances, i.e.

$$C = B + 2(g_1 + R) \qquad \ldots (7.4.2)$$

or, to satisfy the bolt spacing requirement, i.e.

$$C = nB_s/\pi \qquad \ldots (7.4.3)$$

Where, C = bolt-circle diameter,

B = inside diameter of flange,

g_1 = thickness of hub at back of flange,

R = radial clearance from bolt circle to point of connection of hub or nozzle and back of flange,

n = actual number of bolts,

B_s = bolt spacing.

Larger of the two values obtained from Eqs. 7.4.2 and 7.4.3 is to be taken. When both the values become approximately equal, optimum value may be got.

7.5 BOLT MATERIALS

Carbon steel with a guaranteed yield point of not less than 235 MN/m² at room temperature is quite adequate for bolts at working temperatures below 300 °C. Material specifications for bolts and nuts are given in IS : 1363—1967, IS : 1364—1967 and IS : 1862—1967. Alloy steel is to be chosen if service conditions require it and carbon steel is found to be not adequate. Table 7.5 gives allowable stresses for bolting materials in MN/m².

Table 7.5 Allowable Stresses for Bolting Material in MN/m²

Material	Allowable stress MN/m² for design metal temperature not exceeding (°C)						
	50	100	200	250	300	350	400
Hot rolled carbon steel	57.3	55.1	53.5	47.6	—	—	—
5% Cr Mo steel	138.0	138.0	138.0	138.0	138.0	138.0	138.0
15-8 Cr Ni steel	129.0	109.0	85.0	78.5	76.0	73.2	72.0
13% Cr Ni steel	176.0	162.0	140.5	134.0	126.5	119.0	104.5
18 Cr 2 Ni steel	212.0	195.0	170.0	161.0	152.0	144.0	127.0

7.6 FLANGE CALCULATIONS

7.6.1 Flange or bolt loads

The loads acting on the flange are to be calculated both for operating as well as for bolting-up conditions and greater of the two will be taken for determining the flange stresses.

Under the operating conditions it is required to resist the hydrostatic end force of the design pressure tending to part the joint, and to maintain on the gasket or joint-contact surface sufficient compression to assure a leak-tight joint, all at the design temperature. The minimum load under operating conditions is, therefore, a function of the design pressure, the gasket material, and the effective gasket or contact area to be kept tight under pressure.

The bolting-up conditions are required to be considered because the gasket or joint-contact surface is made to be seated by applying an initial load with the bolts when assembling the joints at atmospheric temperature and pressure. The minimum initial load considered to be adequate for proper seating is a function of gasket material and the effective gasket or contact area to be seated.

7.6.1.1 *Determination of bolt loads under internal pressure*

To retain a leak-tight joint under operating condition the minimum bolt load required W_0 is given by:

$$W_0 = H + H_p \qquad \ldots (7.6.1)$$

where, $H = $ load due to design pressure, p, acting on an area $\pi G^2/4$

$= \dfrac{\pi G^2}{4} p$

$= $ total hydrostatic end force,

$H_p = $ load to achieve adequate compression of the gasket under operating condition,

$= \pi G (2b) mp$

G = diameter at location of gasket load reaction
 = mean diameter of gasket contact face, if $b_o \leqslant 6.3$ mm
 = outside diameter of gasket contact face less $2b$, if $b_o > 6.3$ mm.
$(2b)$ = effective gasket pressure width,
b = effective gasket seating width, (Table 7.2)
m = gasket factor, (Table 7.1)

7.6.1.2 Determination of bolt load under bolting-up condition

For initial gasket seating the minimum bolt load required W_g is given by:
$$W_g = \pi Gby \qquad \ldots (7.6.2)$$
Where, y = minimum gasket seating stress, (Table 7.1)

7.6.2 Determination of minimum bolt area theoretically required, A_m

If A_0 is the bolt area required under operating condition and A_g is the area required under bolting-up condition, then,

$$A_0 = \frac{W_0}{S_0} \qquad \ldots (7.6.3)$$

$$A_g = \frac{W_g}{S_g} \qquad \ldots (7.6.4)$$

Where, S_0 = allowable stress for bolting material at design temperature, (Table 7.5)
S_g = allowable stress for bolting material at atmospheric temperature, (Table 7.5)

Theoretically required minimum bolt area, A_m, will be larger of A_0 and A_g. For ideal design A_0 and A_g should be approximated equal.

7.6.2.1 Determination of actual bolt area, A_b

Actual bolt area, A_b, will not be less than A_m to satisfy the theoretical requirement. Next, a standard bolt diameter is to be selected; actual number of bolts should be such that it is a multiple of 4 from practical consideration and also the bolt spacing should not be too large or too small. After satisfying all these requirements A_b becomes usually larger than A_m. It will be economical if the difference is small.

To prevent damage to the gasket during bolting-up, following condition is to be satisfied.

$$\frac{A_b S_g}{\pi G N} < 2y \qquad \ldots (7.6.5)$$

Where, N = actual width of the gasket in contact.
If Eq. 7.6.5 is not satisfied, gasket material should be changed.
It is to be noted that A_b is to be calculated taking root area of the bolts.

7.6.3 Flange stresses

To determine the stresses the flanges are categorized into 3 types, namely, loose-type flanges, integral-type flanges and optional-type flanges.

110   CHEMICAL EQUIPMENT DESIGN—MECHANICAL ASPECTS

Loose-type flanges cover those designs where the method of attachment is not considered to give the mechanical strength equivalent of integral attachment. Fig. 7.3 shows some typical loose-type flanges, the location of loads, and welds.

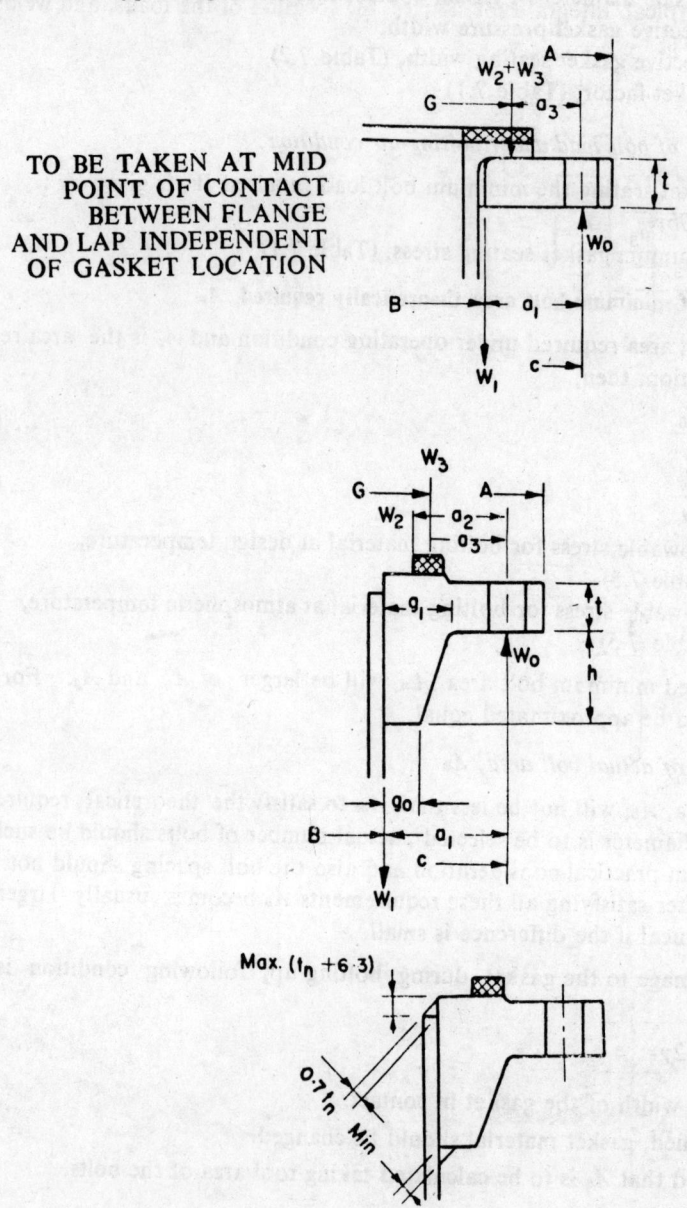

TO BE TAKEN AT MID
POINT OF CONTACT
BETWEEN FLANGE
AND LAP INDEPENDENT
OF GASKET LOCATION

Fig. 7.3 Loose-type flanges

DESIGN OF NON-STANDARD FLANGES

Integral-type flanges cover designs of such a nature that the flange and nozzle neck, vessel, or pipe wall is considered to be the equivalent of an integral structure. In welded construction, the nozzle neck, vessel, or pipe wall is considered to act as a hub. Fig. 7.4 shows some typical integral-type flanges, the location of the loads, and welds and other construction details.

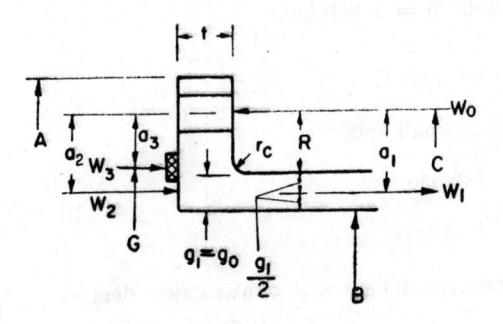

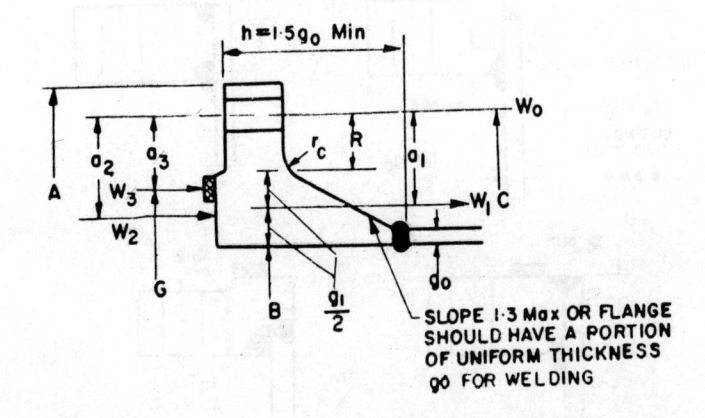

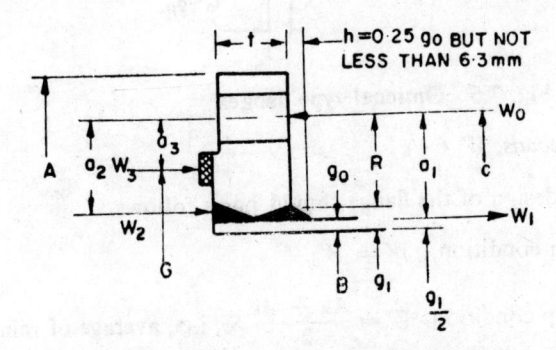

Fig. 7.4 Integral-type flanges

112   CHEMICAL EQUIPMENT DESIGN—MECHANICAL ASPECTS

Optional-type flanges cover designs where the attachment of the flange to the nozzle neck, vessel, or pipe wall is such that the assembly is considered to act as a unit, which should be calculated as an integral flange. But for simplicity one can calculate the construction as a loose-type flange provided none of the following values are exceeded:

$$g_0 = 16 \text{ mm}; \quad \frac{B}{g_0} = 300; \quad p = 2 \text{ MN/m}^2,$$

operating temperature = 365 °C

Where, g_0 = thickness of hub at small end,

B = inside diameter of flange,

p = design pressure.

Fig. 7.5 shows some typical optional-type flanges with construction details.

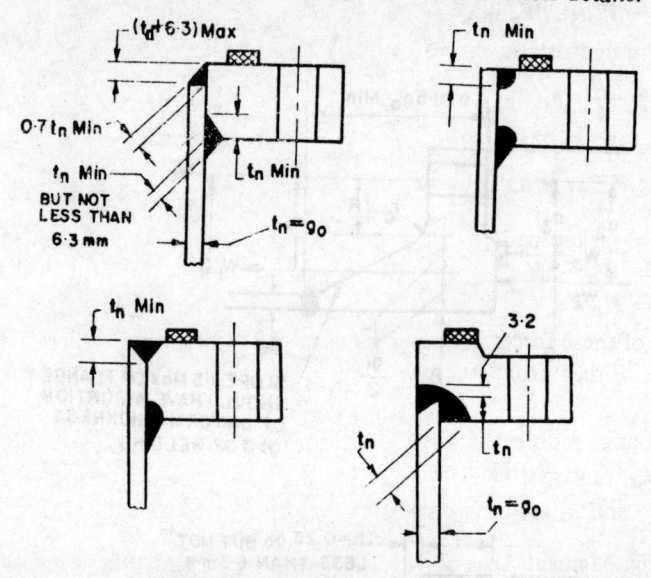

Fig. 7.5 Optional-type flanges

7.6.3.1 Design bolt or flange loads, W

The loads used in the design of the flange should be as follows:

For operating condition, $W = W_0$

For bolting-up condition, $W = \dfrac{A_m + A_b}{2} S_g$, i.e., average of minimum and maximum bolt loads.

DESIGN OF NON-STANDARD FLANGES

7.6.3.2 Determination of flange moments

Flange moments are to be calculated for both the operating and the bolting-up conditions. Larger of the two, after multiplying with bolt pitch correction factor, C_F, is to be used for determining the flange stresses. This correction factor is introduced for
$$B_s \neq 2d + t$$
and is given by
$$C_F = \sqrt{\frac{\text{actual bolt spacing}}{2d + t}}$$
For initial calculation C_F may be taken as unity and finally its magnitude can be checked.

(a) Operating condition :

Under operating condition the load W_0 can be said comprising of 3 load components, such as,
$$W_0 = W_1 + W_2 + W_3 \qquad \ldots (7.6.6)$$
Where, $\quad W_1 =$ hydrostatic end force on area inside of flange,
$$= \frac{\pi B^2}{4} p.$$
$$W_2 = H - W_1$$
$$= \frac{\pi}{4}(G^2 - B^2) p$$
$W_3 =$ gasket load
$$= W_0 - H$$
$$= H_p$$

Locations of these forces acting in the flange are shown in Figs. 7.3 and 7.4. The distances of these forces from the bolt circle diameter give the respective moment arms a_1, a_2, and a_3.

The total flange moment is, therefore, given by :
$$M_0 = W_1 a_1 + W_2 a_2 + W_3 a_3 \qquad \ldots (7.6.7)$$
The values of a_1, a_2 and a_3 for different flange types are given in Table 7.6.

Table 7.6 Moment Arms for Flange Loads under Operating Conditions

Type of flanges	a_1	a_2	a_3
Integral type flanges (Fig. 7.4)	$R + \dfrac{g_1}{2}$	$\dfrac{R + g_1 + a_3}{2}$	$\dfrac{C - G}{2}$
Loose type except lap joint flanges (Fig. 7.3)	$\dfrac{C - B}{2}$	$\dfrac{a_1 + a_3}{2}$	$\dfrac{C - G}{2}$
Lap joint flanges (Fig. 7.3)	$\dfrac{C - B}{2}$	$\dfrac{C - G}{2}$	$\dfrac{C - G}{2}$

(b) Bolting-up condition :

In this case the total flange moment is given by :

$$M_g = W a_3 \qquad \ldots (7.6.8)$$

where, $W = \dfrac{A_m + A_b}{2} S_g$ and $a_3 = \dfrac{C - G}{2}$

7.6.3.3. Calculation of induced flange stresses

For this purpose, design flange moment $M = M_0$ or (ϕM_g), whichever is larger, is to be taken.

Here, ϕ is the ratio of the allowable stresses of the flange materials at design and atmospheric temperatures, i.e., $\phi \leqslant 1$.

(a) For integral-type flanges the correlations for flange stresses are given below[2] :

S_Z = axial or longitudinal bending stress in the hub or nozzle wall (due to discontinuity at the junction of flange ring and hub or nozzle),

$$= X f \dfrac{M C_F}{B g_1^2} \qquad \ldots (7.6.9)$$

S_R = radial stress on the inside surface of the flange ring (at the junction with the hub),

$$= \left(1 + 1.33 F \dfrac{1}{\sqrt{B g_0}}\right) \dfrac{M C_F}{B t^2} X \qquad \ldots (7.6.10)$$

S_T = tangential stress in the flange ring (at the same point as of S_R)

$$= \dfrac{M C_F}{B t^2} Y - Z S_R \qquad \ldots (7.6.11)$$

The coefficients X, Y, Z are calculated from the following equations :

$$X = \dfrac{1}{\dfrac{1}{T}\left(1 + \dfrac{tF}{\sqrt{B g_0}}\right) + \dfrac{V t^3}{U g_0^2 \sqrt{B g_0}}} \ ;$$

$$Y = \dfrac{0.955}{K-1}\left[(1 - \mu) + (1 + \mu)\, 4.605\, \dfrac{K^2 \log K}{K^2 - 1}\right]$$

$$Z = \dfrac{K^2 + 1}{K^2 - 1} \ ;$$

where, $K = \dfrac{A}{B}$

$$T = 0.955\, \dfrac{K^2 \left(1 + 4.605\, \dfrac{1 + \mu}{1 - \mu} \log K\right) - 1}{(K - 1)\left(1 + \dfrac{1 + \mu}{1 - \mu} K^2\right)} \ ;$$

$$U = \dfrac{0.955 K^2 \left(1 + 4.605\, \dfrac{1 + \mu}{1 - \mu} \log K\right) - 1}{(1 + \mu)(K - 1)(K^2 - 1)}$$

DESIGN OF NON-STANDARD FLANGES

Usually flanges are made with steel for which Poisson's ratio $\mu = 0.3$. The values of Y, Z, T and U are presented graphically in Fig 7.6 for steel flanges.

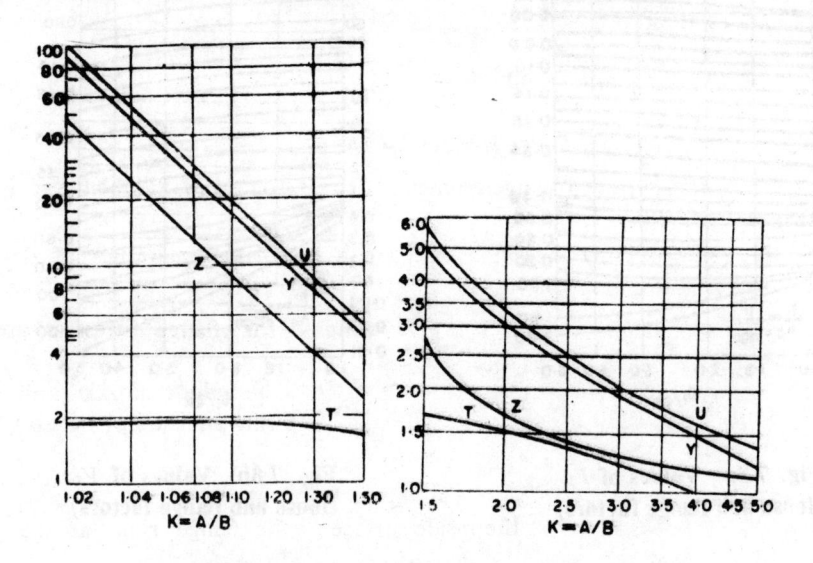

Fig. 7.6 Values of T, U, Y and Z for $K = A/B$

The coefficients F and V as functions of $\dfrac{h}{\sqrt{B\,g_0}}$ and $\dfrac{g_1}{g_0}$ are presented in Fig. 7.7 for integral-type flanges and in Fig. 7.8 for loose-type flanges (F_L and V_L).

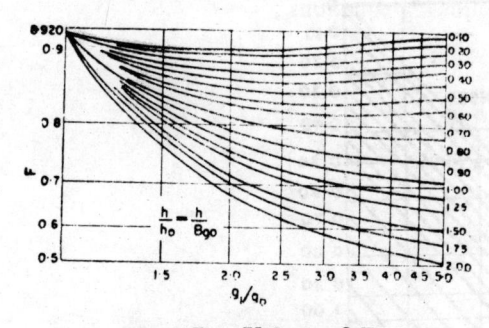

Fig. 7.7a Values of F (integral flange factors)

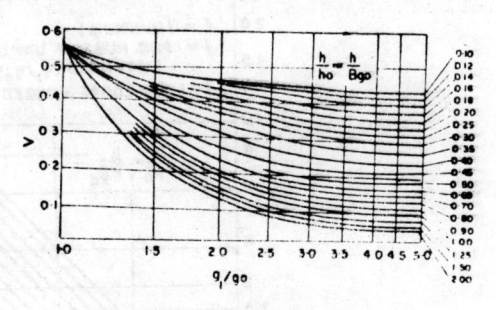

Fig. 7.7b Values of V (integral flange factors)

The coefficient f is obtained from Fig. 7.9 as functions of $\dfrac{h}{\sqrt{B\,g_0}}$ and $\dfrac{g_1}{g_0}$. Minimum value of $f = 1$ for all cases. For flanges with hub of uniform thickness $\left(\dfrac{g_1}{g_0} = 1\right)$ and for loose-type flanges, $f = 1$.

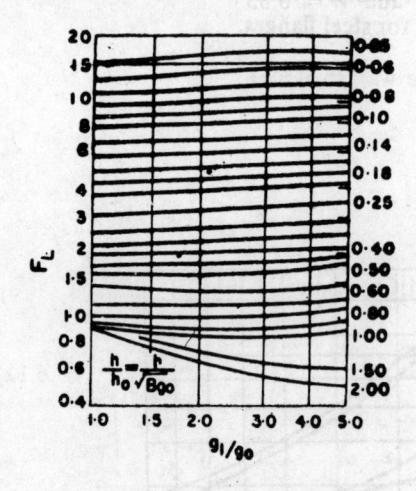

CHEMICAL EQUIPMENT DESIGN—MECHANICAL ASPECTS

Fig. 7.8a Values of F_L
(loose hub flange factors)

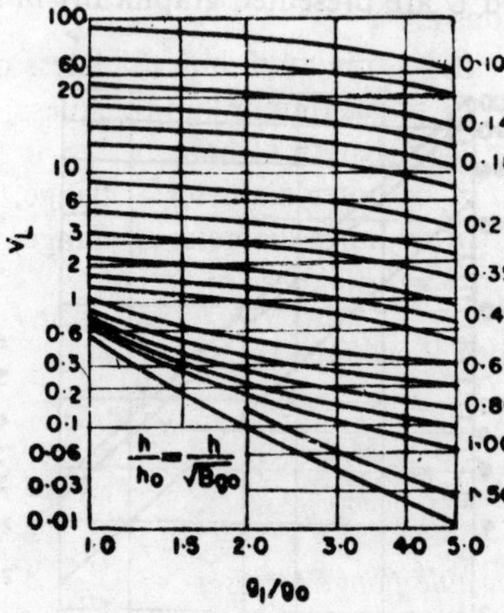

Fig. 7.8b Values of V_L
(loose hub flange factors)

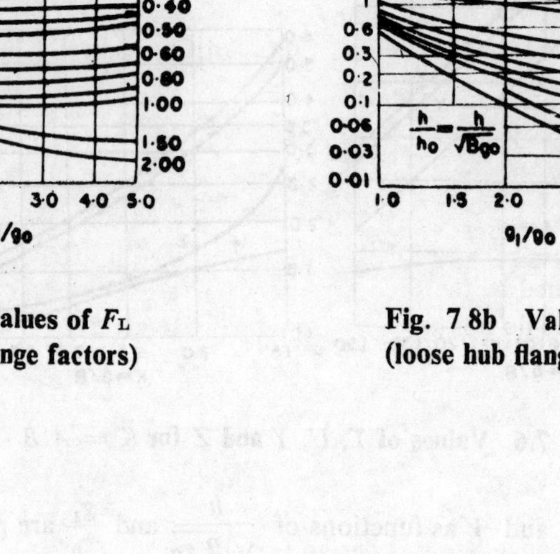

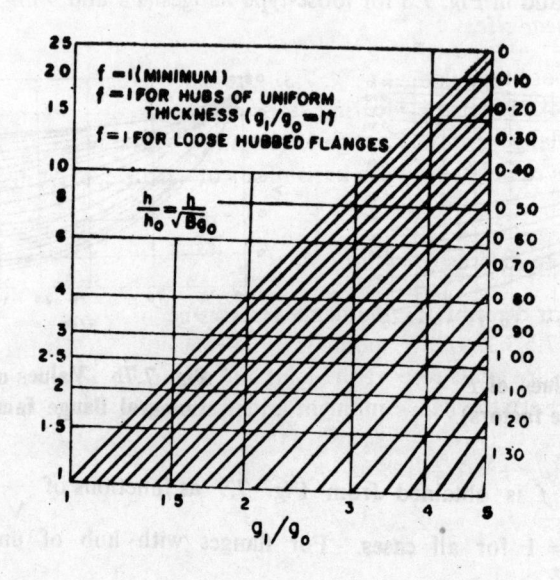

Fig. 7.9 Values of f (hub stress correction factor)

DESIGN OF NON-STANDARD FLANGES

Again, for the case of $\frac{g_1}{g_0} = 1$; $F = 0.91$ and $V = 0.55$

Other notations are:

g_0 = minimum hub thickness or nozzle wall thickness,
g_1 = maximum hub thickness,
h = length of hub,
A = outer diameter of flange,
B = inner diameter of flange,
t = thickness of flange.

(b) For loose-type ring flanges following relations apply:

$S_R = S_Z = 0$

$S_T = \dfrac{M\, C_F}{B\, t^2}\, Y$ ... (7.6.12)

7.6.3.4 Allowable flange stresses[3]

The flanges designed by the procedure given in section 7.6.3.3 can be considered safe, if the induced stresses do not exceed the following limits:

$S_Z = 1.5\, S_{FO}$
$S_R = S_{FO}$
$S_T = S_{FO}$
$S_Z + S_R = 2\, S_{FO}$
$S_Z + S_T = 2\, S_{FO}$

where, S_{FO} = allowable flange stress at design temperature.

7.6.3.5 Checking for shear stress

In the case of loose-type flanges (Fig. 7.3), the lap or weld may be subjected to shear. It is to be checked that the shearing stress carried by lap or welds does not exceed 0.8 times the allowable flange stresses for gasket seating and operating conditions. The shearing stress should be calculated on the basis of H_p or W_g as defined in paragraph 7.6.1, whichever is larger.

7.6.4 Flanges subjected to external pressure

The procedure for determining the flange stresses under external pressure will be the same as given in 7.6.3.3. Only point is to be remembered that the hydrostatic end force under internal pressure tends to part the joint and under external pressure the effect is just opposite. As a result the flange moment under external pressure is slightly different and is given by,

$M_o = W_1 (a_1 - a_3) + W_2 (a_2 - a_3)$ ... (7.6.13)

where $W_1 = \dfrac{\pi B^2}{4}\, p_e,$

[3] IS: 2825—1969.

$$W_2 = H - W_1$$

$$H = \frac{\pi G^2}{4} p_e$$

Here, P_e = external design pressure.

For gasket seating,

$$M_g = W\, a_3$$

where, $W = \dfrac{A_g + A_b}{2} S_g.$ ... (7.6.14)

Rest of the calculations are same as for flanges under internal pressure.

7.6.5. A few remarks on flange design

If the correlations given for flange stress calculation in 7.6.3.3 is analysed, following observations can be made and this may help the students and inexperienced designer to make logical decisions while designing the flange connections. Basically two question may trouble their mind.

(a) When to decide for loose-type and when for integral-type flanges ?

(b) How to choose ring only type or hubbed type integral flanges ?

The utility and service life of different types of flanges are discussed in section 7.1. Fatigue life is very poor for loose-type flanges. For high pressure service as well at high temperature loose-type flanges will not be economical. The reasons are :

(1) Loose-type flanges are subjected to tangential stress (S_T) only. From Eq. 7.6.12 it is seen that the induced stress can be brought down to the allowable limit by increasing flange thickness. Therefore, if pressure is high, flange moment will be large and if temperature is high, allowable flange stresses will be low. As a result a massive flange will be required which may be quite costly.

(2) Welds may fail due to shear.

(3) Thermal stresses may develop due to improper contact between flange and nozzle.

If it is decided to go with the integral-type flanges, the following points are to be considered :

(1) Ring only type flange is easier to construct and generally cheaper if weight variation is not too large.

(2) With the increase of flange thickness (t), all the induced stresses are reduced.

(3) If radial stress (S_R) and tangential stress S_T are controlling, ring only type integral flange can be selected.

(4) When hub stress (S_z) is predominant, hubbed-type integral flange may be found suitable. It may be mentioned here that this longitudinal bending stress in the hub or nozzle is caused due to discontinuity at the junction of hub and flange ring. The analysis is available in the literature.[4,5]

(5) Hub length (h) is a function of damping factor (β). In section 5.2.5 it is observed that the length of cylindrical cross-section affected by load deformation can go upto π / β but most effective length is $1 / \beta$, where, $\beta = 1.82 / \sqrt{B g_0}$. Therefore, a hub length of $\sqrt{B g_0}/1.82$ or less can be selected. Larger value of hub length reduces the hub stress correction factor (f), but increases the coefficient, X, in Eq. 7.6.9. Hence a compromise is to be arrived at. It may be economical to keep hub length smaller.

(6) The ratio of outer flange diameter to inner diameter influences the induced flange stresses. It is economical to select outer flange diameter (A) as low as practicable.

(7) Weldment at the junction between flange ring and nozzle is also considered as hub.

7.6.6 Some standard dimensions[6]

This section is introduced to inform the designer how one can select various reasonable dimensions while going with the stress analysis. These are only for guidance. Higher or lower values can be taken if so desired.

(1) For loose-type and lap-joint flanges the inner flange diameter (B) is 2-3 mm larger than the nominal or outer diameter of the nozzle or shell.

(2) For integral type flanges, B = inner diameter of shell or nozzle.

(3) Minimum hub thickness (g_0) is equal to the shell or nozzle wall thickness.

(4) The depth of the groove for 'tongue and groove' and 'male-female' types is normally 5 mm and tongue height is 6 mm.

(5) The difference in width between tongue and groove is 2 mm.

(6) The minimum width of the tongue is 10 mm. If the bolt load or flange diameter is large, the width may be increased by 3 mm upto 16 mm.

(7) The clearance between tongue and groove of 'male-female' type is 1 mm.

(8) Nominal (or outer) diameters (in mm) of nozzle or shells given for standard flanges are[6] :

324, 368, 419 ; (457), 508, 600 ; 700, 800, 900 ; 1 000, 1 100, 1 200 ; (1 300), 1 400, (1 500) ; 1 600, (1 700), 1 800 ; (1 900), 2 000, (2 100) ; 2 200, (2 300), 2 400 ; 2 600, 2 800, 3 000 ; 3 200, (3 400), 3 600 ; 3 800, 4 000.

(Sizes shown in the parentheses are of second preference)

---

[4] John F. Harvey, "Pressure Vessel Design", East-West Press Pvt. Ltd., New Delhi.
[5] L. E. Brownell and E. H. Young, "Process Equipment Design", John Wiley and Sons, Inc., New York.
[6] IS : 4864—1968 to IS : 4870—1968.

(9) Standard flange thickness (t) in mm are :

20, 25, 30, 35, 40, 45, 50, 55, 60, 65, 70,
75, 80, 85, 90, 95, 100, 105, 110, 120.

(10) The number of bolts specified for nozzle diameter 324 mm is 12 and bolt diameter M 16.

(11) Bolt diameters (d) used for standard flanges are :

M 16 and M 20 for loose-type flanges.
M 16, M 20, M 24, M 27, M 30 for integral type flanges.

(12) Maximum hub thickness (g_1) is approximately 0.5 times the flange thickness.

(13) Hub length (h) is approximately calculated from the relationship obtained from damping factor of discontinuity stresses. Standard values given in mm are :

20, 25, 30, 35, 40, 45, 50, 55, 60, 65.

(14) Branch pipes or nozzle :

The thickness of the branch pipe or nozzle should be adequate to meet the design requirement and in addition it should take into consideration corrosion, erosion, loads transmitted from connecting piping, etc. But in no case the thickness can be taken less than the values given in Table 7.7.

Table 7.7 Minimum Nozzle Thickness (IS : 2825—1969)

Branch Nominal size (mm)	Minimum corroded thickness (mm)	
	Carbon & Ferritic Alloy Steels	Austenitic Stainless Steel
51 and smaller	5	3
65, 80, 90	6	5
100, 150	8	6
203, 254	10	8
305	11	10
356	13	10
406, 457	16	13

Design Example 7.1 : A ring-type flange with a plain face for a heat exchanger shell is required to be designed to the following specifications :

Design pressure = 1 MN/m^2
Design temperature = 150 °C

DESIGN OF NON-STANDARD FLANGES

Flange material = IS : 2004 — 1962 Class 2
Bolting steel = 5% Cr Mo steel
Gasket material = asbestos composition
Shell outside diameter = 1 m = B
Shell thickness = 0.01 m = g_0
Shell inside diameter = 0.98 m
Allowable stress of flange material = 100 MN/m²
Allowable stress of bolting material = 138 MN/m²
Flange type = optional (Fig. 7.5)

Solution : Determination of gasket width by Eq. 7.3.2

$$\frac{d_o}{d_i} = \left(\frac{y - pm}{y - p(m+1)}\right)^{\frac{1}{2}}$$

Assuming a gasket thickness of 1.6 mm, from Table 7.1 :

$$y = 25.5$$
$$m = 2.75$$

Substituting,

$$\frac{d_o}{d_i} = \left(\frac{25.5 - 1 \times 2.75}{25.5 - 1(2.75 + 1)}\right)^{\frac{1}{2}}$$
$$= 1.021$$

Let d_i of the gasket equal 1.01 m, i.e., 10 mm larger than B, then

$$d_o = (1.021)(1.01) = 1.032 \text{ m}$$

Minimum gasket width $= \left(\frac{1.032 - 1.01}{2}\right) = 0.011$ m

Hence, a gasket width of 0.012 m is selected, and d_o becomes 1.034 m

Basic gasket seating width, $b_o = \frac{12}{2} = 6.0$ mm

Diameter at location of gasket load reaction is $G = d_i + N$
$$= 1.01 + 0.012$$
$$= 1.022 \text{ m}$$

Estimation of bolt loads :

Load due to design pressure,

$$H = \frac{\pi G^2}{4} p$$
$$= \frac{\pi (1.022)^2}{4} \quad (1)$$
$$= 0.825 \text{ MN}$$

Load to keep joint tight under operation
$$H_p = \pi G (2b) mp$$
$$= \pi (1.022)(0.012)(2.75)(1)$$
$$= 0.106 \text{ MN}$$

Total operating load (Eq. 7.6.1)
$$W_o = H + H_p$$
$$= 0.825 + 0.106$$
$$= 0.931 \text{ MN}$$

Load to seat gasket under bolting-up condition (Eq. 7.6.2)
$$W_g = \pi G b y$$
$$= \pi (1.022)(0.006)(25.5)$$
$$= 0.495 \text{ MN}$$

W_0 is larger than W_g and therefore, controlling load $= 0.931$ MN

Calculation of minimum bolting area (Eq. 7.6.3)
$$A_m = A_0 = \frac{W_0}{S_0} = \frac{0.931}{138}$$
$$= 6.75 \times 10^{-3} \text{ m}^2$$

Calculation of optimum bolt size (Table 7.4)

In this case $g_1 = \dfrac{g_0}{0.707} = 1.415 g_0$ for weld leg.

Bolt size	Root area (m^2)	Min No. of bolts	Actual No. of bolts (n)	R (m)	B_s (m)	$C = \dfrac{nB_s}{\pi}$ (m)	$C = ID + 2 \times (1.415 g_0 + R)$ (m)
M 16 × 1.5	1.33×10^{-4}	50.8	52	0.025	0.075	1.24	1.058 3
M 18 × 2	1.54×10^{-4}	43.7	44	0.027	0.075	1.05	1.062 3
M 20 × 2	2.00×10^{-4}	33.7	36	0.030	0.075	0.86	1.068 3

From the above calculation the minimum bolt-circle is 1.062 3 m when using M 18 bolts. For simplicity in dimensioning 44 bolts of 18 mm diameter on a 1.07 m bolt-circle are specified.

Bolt-circle diameter, $C = 1.07$ m

Calculation of flange outside diameter (A):
$$A = C + \text{bolt diameter} + 0.02 \text{ m (minimum)}$$
$$= 1.07 + 0.018 + 0.02$$
$$= 1.108 \text{ m}$$
$$= 1.11 \text{ m (selected)}.$$

DESIGN OF NON-STANDARD FLANGES

Check of gasket width (Eq. 7.6.5)

$$\frac{A_b S_a}{\pi G N} = \frac{(44 \times 1.54 \times 10^{-4})(138)}{\pi (1.022)(0.012)}$$

$$= 24.2 < 2y$$

Condition is satisfied.

Flange moment computations:

(a) For operating condition:

From Eq. 7.6.6,

$$W_0 = W_1 + W_2 + W_3$$

Where,
$$W_1 = \frac{\pi B^2}{4} p$$
$$= 0.785 (1)^2 (1)$$
$$= 0.785 \text{ MN}$$
$$W_2 = H - W_1$$
$$= 0.825 - 0.785 = 0.04 \text{ MN}$$
$$W_3 = W_0 - H$$
$$= H_p = 0.106 \text{ MN}$$

From Eq. 7.6.7
$$M_0 = W_1 a_1 + W_2 a_2 + W_3 a_3$$

From Table 7.6
$$a_1 = \frac{C - B}{2}$$
$$= \frac{1.07 - 1.00}{2} = 0.035 \text{ m}$$
$$a_3 = \frac{C - G}{2}$$
$$= \frac{1.07 - 1.022}{2} = 0.024 \text{ m}$$
$$a_2 = \frac{a_1 + a_3}{2}$$
$$= \frac{0.035 + 0.024}{2} = 0.030 \text{ m}$$

Therefore,
$$M_0 = (0.785)(0.035) + (0.04)(0.03) + (0.106)(0.024)$$
$$= 2.75 \times 10^{-2} + 0.12 \times 10^{-2} + 0.255 \times 10^{-2}$$
$$= 3.125 \times 10^{-2} \text{ MJ}$$

(b) For bolting-up condition (no internal pressure):
From Eq. 7.6.8
$$M_g = W a_3$$
Where,
$$W = \frac{A_m + A_b}{2} S_g$$
$$A_m = 6.75 \times 10^{-3} \text{ m}^2$$
$$A_b = 44 (1.54 \times 10^{-4})$$
$$= 6.76 \times 10^{-3} \text{ m}^2$$
$$S_g = 138 \text{ MN/m}^2$$
$$W = \frac{(6.75 + 6.76)(10^{-3})(138)}{2}$$
$$= 0.932 \text{ MN}$$
$$a_3 = 0.024 \text{ m}$$
$$M_g = (0.932)(0.024)$$
$$= 2.24 \times 10^{-2} \text{ MJ}$$

M_0 is greater than M_g. Hence moment under operating condition (M_0) is controlling.
$$M = M_0 = 3.125 \times 10^{-2} \text{ MJ}.$$

Calculation of flange thickness:
From Eq. 7.6.12
$$t^2 = \frac{M C_F Y}{B S_T} = \frac{M C_F Y}{B S_{FO}}$$
$$K = \frac{A}{B} = \frac{1.11}{1} = 1.11$$

From Fig. 7.6 with K equals 1.11
$$Y = 18.55$$
Therefore, assuming $C_F = 1$
$$t^2 = \frac{(3.125 \times 10^{-2})(18.55)}{(1)(100)} = 5.8 \times 10^{-3} \text{ m}^2$$
or
$$t = 0.076 \text{ m}$$

Actual bolt spacing $(B_s) = \frac{\pi C}{n}$
$$= \frac{\pi (1.07)}{44}$$
$$= 0.076\ 5 \text{ m}$$

Bolt pitch correction factor (C_F)
$$C_F = \left(\frac{B_s}{2d + t}\right)^{\frac{1}{2}}$$
$$= \left(\frac{0.076\ 5}{0.036 + 0.076}\right)^{\frac{1}{2}}$$
$$= 0.83$$
$$\sqrt{C_F} = 0.912$$

DESIGN OF NON-STANDARD FLANGES

The flange thickness calculated above is to be multiplied by $\sqrt{C_F}$. Hence, actual flange thickness equals to

$$t = 0.076 \times 0.912$$
$$= 0.069 \text{ m}$$

Select a flange thickness of 70 mm.

It can be observed from Eq. 7.6.12 and Fig. 7.6, with the increase of factor K i.e., with the increase of outside flange diameter A, the coefficient Y decreases, that means, flange thickness also decreases. Let it be examined, what should be the outer flange diameter, to bring down the flange thickness to a value, when C_F becomes unity.

From the correlation, C_F becomes approximately unity, if $t = 40$ mm

From Eq. 7.6.12

$$Y = \frac{Bt^2 S_{FO}}{M}$$

$$= \frac{(1)(1.6 \times 10^{-3})(100)}{3.125 \times 10^{-2}} = 5.1$$

and $K = 1.48$

That means, a flange outer diameter of 1.48 m for a shell diameter of 1 m is required and this is neither a good proportion, nor economical.

CHAPTER 8

DESIGN OF PROCESS VESSELS AND PIPES UNDER EXTERNAL PRESSURE

8.1 INTRODUCTION

Many of the chemical process equipment are required to be operated under such condition, when the inside pressure is lower than the outside pressure. This may be due to inside vacuum or outside higher pressure or combination of both (see paragraph 2.3) A few examples of such cases are : multiple-effect evaporator and condenser for last effect ; vacuum distillation column and crystallizers ; jacketed vessels ; etc.

Because of external pressure effects the cylindrical vessels experience an induced circumferential compressive stress equal to twice the longitudinal compressive stress. As a result the vessel is apt to fail because of elastic instability caused by the circumferential compressive stress. The rigidity of the vessels, under such condition may be increased by using uniformly spaced, internal or external circumferential stiffening rings (Fig. 8.1). This reduces the effective length of the vessels to the centre-to-centre distance of the stiffeners. L/D_o (length to diameter ratio) is, therefore, a significant parameter in determining the safe pressure under external compressive stress.

If the vessel is deformed to an extent, when normal operation cannot be carried out with it for which it is specified, it may be considered that the vessel has failed, though no bursting occurred. Under external pressure the vessels are subjected to two kinds of failure. These are due to :

(1) Elastic instability or buckling within proportional limit. This occurs with the cylinders having effective length larger than the "critical length". The corresponding critical pressure at which buckling occurs is a function of the t/D_o ratio and the modulus of elasticity, E, of the material at design temperature. Geometrical irregularities like lobes in the shell cause buckling at lower pressure.

(2) A vessel of moderate thickness may collapse under external pressure at stresses above the proportional limit but below the yield point (i.e., $\dfrac{p D_o}{2 t} >$ proportional limit).

If the length of the shell with closures, L, or distance between circumferential stiffeners, L, as the case may be, is less than the critical length, the critical pressure at which collapse occurs is a function of the L/D_o ratio as well as of the t/D_o ratio and the modulus of elasticity, E. Out of roundness may cause failure at lower critical pressure.

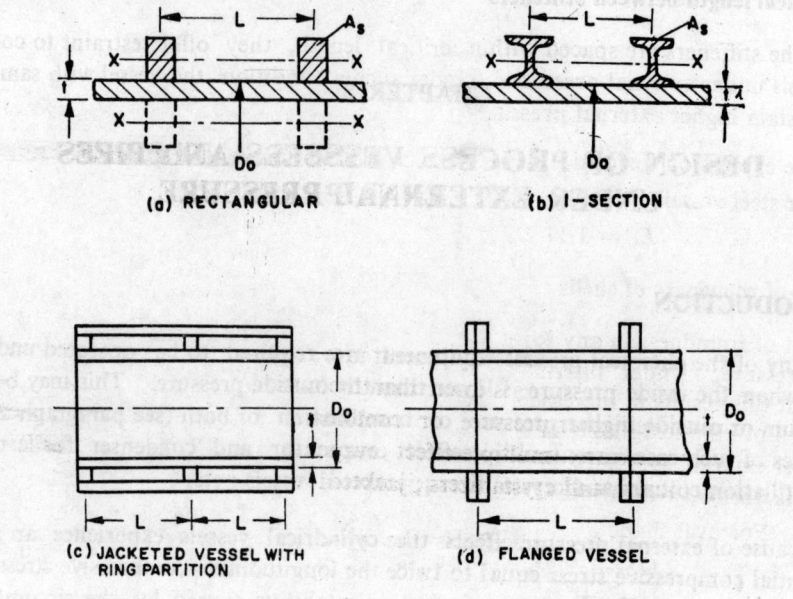

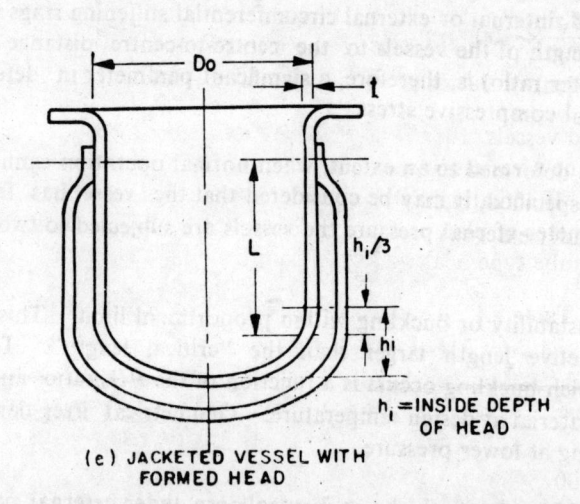

Fig. 8.1 Reinforced vessel under external pressure showing effective length and cross sectional area of stiffening rings

Proportional limit is defined as the greatest stress which a material can sustain without deviating from the law of stress-strain proportionality (i.e., Hooke's law).

DESIGN OF PROCESS VESSELS AND PIPES UNDER EXTERNAL PRESSURE

8.1.1 Critical length between stiffeners

If the stiffeners are spaced within critical length, they offer restraint to collapsing of the vessels under external pressure. Under such a condition the vessel with same thickness can sustain higher external pressure.

The expression for critical length is available in literature as given below.[1]

For steel vessels ($\mu = 0.3$)

$$L_c = 1.11 \, D_o \, \sqrt{D_o/t} \qquad \ldots (8.1.1)$$

8.1.2 Out-of-roundness of shells

Out-of-roundness in any form is very much detrimental to the vessel strength under external pressure. Under internal pressure out-of-roundness does not cause so much worries, but this results in increased stress concentration under external pressure. As a result a shell of elliptical shape, or a circular shell, either dented or with flat spots, is less strong under external pressure than a vessel having a true cylindrical shape. Out-of-roundness factor, U, is determined as follows :

(a) For oval shape :

$$U = \frac{2\,(D_{max} - D_{min})}{D_{max} + D_{min}} \times 100, \text{ per cent}$$

(b) For dent or flat spots :

$$U = \frac{4\,a}{D_o} \times 100, \text{ per cent}$$

Where, $a =$ depth of dent or flat spots (maximum value is to be taken).

In case of old vessels, the largest value from (a) and (b) is to be taken for design calculation. For the new vessels, whose out-of-roundness is not known, $U = 15\%$ (minimum) is to be taken. If actually measured, smaller values can be used. In this connection the tolerances on diameter of plate shells prescribed in Indian Standard (IS : 4503 — Specification for shell and tube type heat exchanger) is shown in Table 8.1.

Table 8.1 Tolerances on Diameter of Plate Shells
(All dimensions in mm)

Nominal Diameter	Permissible Deviation	
	Grade I	Grade II
200 upto 400	± 3	—
Over 400 upto 600	± 3	+ 6 − 3
Over 600 upto 800	± 4	+ 7 − 4
Over 800 upto 1 000	± 5	+ 8 − 5
Over 1 000	± 6	+ 8 − 6

[1] L. E. Brownell and E. H. Young, "Process Equipment Design", John Wiley and Sons, Inc., New York.

Grade I implies that greater care should be taken to ensure that no distortion occurs when fittings are welded on.

8.2 DETERMINATION OF SAFE PRESSURE AGAINST ELASTIC FAILURE

A cylindrical shell under external pressure tends to deform inward as a result of the external radial pressure. The relationship between the radius of curvature, the product EI and bending moment producing curvature for long thin cylinders under external pressure is given by Timoshenko.[2]

$$M = \frac{EI}{r} \qquad \ldots (8.2.1)$$

From Eq. 8.2.1 the theoretical or critical load per unit circumferential length of unit width of circumference is given by:

$$p_c = \frac{3EI}{r_o^3} = \frac{24EI}{D_o^3} \qquad \ldots (8.2.2)$$

For a strip of unit width the critical load is the pressure at which buckling occurs theoretically. This expression assumes no restriction to the deformation from the adjacent metal or ends. To allow for this restraint Eq. 8.2.2 is to be divided by $(1 - \mu^2)$. To express the critical pressure or stress in terms of the shell thickness, t, a substitute for I may be made for a rectangular strip.

$$I = \frac{bt^3}{12}$$

Where, $b = 1$ for a strip of unit width. Substituting in Eq. 8.2.2 gives:

$$p_c = \frac{2E}{1 - \mu^2} \left(\frac{t}{D_o}\right)^3 \qquad \ldots (8.2.3)$$

For steel, Poisson's ratio, $\mu = 0.3$

Therefore,

$$p_c = 2.2 E (t/D_o)^3 \qquad \ldots (8.2.4)$$

The above equation gives the theoretical "critical" external pressure at which a long cylindrical vessel will buckle. From actual investigation the magnitude of p_c was found to be much less than what is given by Eq. 8.2.4. In this expression the vessel is assumed to be compressed radially without any local distortion like lobe formation. Lobes are actually formed in buckling and elastic stability is disturbed at lower critical pressure. Taking into consideration the effect of number of lobes (n) formed in buckling on critical pressure for a circumferential to axial length ratio $Z \left(= \frac{\pi D_o}{L} \right)$, the theoretical expression for critical pressure is available in literature[3] and is given below:

---

[2] S. Timoshenko, "Theory of Elastic Stability", McGraw-Hill, New York.
[3] H. Meincke, "Berechnung und Konstruktion Zylindrischer Behaelter unter Aussendruck", Konstruktion Bd 11 (1959) H. 4 S. 131/137.

DESIGN OF PROCESS VESSELS AND PIPES UNDER EXTERNAL PRESSURE

$$p_c = E \left\{ \frac{2}{(n^2 - 1) [1 + (2n/Z)^2]^2} \left(\frac{t}{D_o} \right) \right.$$
$$\left. + \frac{2}{3(1 - \mu^2)} \left[n^2 - 1 + \frac{2n^2 - 1 - \mu}{(2n/Z)^2 - 1} \right] \left(\frac{t}{D_o} \right)^3 \right\} \quad \ldots (8.2.5)$$

The safe external pressure, p, against elastic failure is obtained from Eq 8.2.5 by dividing p_c by a factor of safety, f_E. According to German code (AD—Merkblaetter) $f_E = 3$ is satisfactory.

To determine $p = \dfrac{p_c}{f_E}$ from Eq. 8.2.5 for vessels of a given $\dfrac{t}{D_o}$ ratio, a group of curves are to be plotted, one curve for each integral value of $n = 2$ or over, with D_o/L as ordinates and p as abscissa. That curve of the group which gives the least value of p is then used to find the p corresponding to a given D_o/L ratio.

Determination of safe external pressure, p, from Eq. 8.2.5 is very time consuming. From the results of this equation, following simplified correlation is suggested for $\mu = 0.3$.

$$p = KE (t/D_o)^m \quad \ldots (8.2.6)$$

Where, p = safe external pressure,
 E = modulus of elasticity at design temperature,
 t = corroded thickness of the vessel,
 D_o = outer diameter of the shell.

The values of K and m as a function of D_o/L ratio is given in Table 8.2.

Table 8.2 Values of K and m in Eq 8.2.6

D_o/L	K	m
0	0.733	3.00
0.1	0.185	2.60
0.2	0.224	2.54
0.3	0.229	2.47
0.4	0.246	2.43
0.6	0.516	2.49
0.8	0.660	2.48
1.0	0.879	2.49
1.5	1.572	2.52
2.0	2.364	2.54
3.0	5.144	2.61
4.0	9.037	2.62
5.0	10.359	2.58

8.3 DETERMINATION OF SAFE EXTERNAL PRESSURE AGAINST PLASTIC DEFORMATION

When the circumferential stiffeners are spaced at less than critical length, the collapsing pressure of the vessel shells increases as a result of the improved rigidity. The coefficient in Eq. 8.2.4 is therefore, to be modified for the new situation and the correlation can be expressed as follows :

$$P_c = K' E (t/D_o)^3 \qquad \ldots (8.3.1)$$

Where, K' = coefficient depends on D_o/L and t/D_o available in literature[1],
$\geqslant 2.22$

The equations derived for internal pressure condition using membrane stress theory (Ch. 3) will apply equally to cases of external pressure with effective length less than critical length and the resulting stresses will be compression. Therefore,

$$f_c = \frac{p_c D_o}{2 t} \qquad \ldots (8.3.2)$$

Where, f_c = Compressive yield stress of the material.

If f_c in Eq. 8.3.2 is substituted by allowable compressive stress, f, the expression for safe external pressure against plastic deformation can be given as ;

$$p = 2 f (t/D_o) \qquad \ldots (8.3.3)$$

In section 8.1.2 it is mentioned that out-of-roundness of the vessel reduces its collapsing strength against external pressure and this defect is inherent with all the vessels. This necessitates the modification of Eq. 8.3.3, which assumes perfect roundness of the vessel, to accommodate out-of-roundness in determining safe external pressure. AD-Merkblaetter (German code) gives the following expression for determining the safe external pressure against plastic deformation considering out-of-roundness.

$$p = 2f \left(\frac{t}{D_o}\right) \frac{1}{1 + \frac{1.5 \, U (1 - 0.2 \, D_o/L)}{100 \left(\frac{t}{D_o}\right)}} \qquad \ldots (8.3.4)$$

Where, t = corroded shell thickness,
U = out-of-roundness in %

Eq. 8.3.4 is applicable for $D_o/L \leqslant 5$, or $L \geqslant 0.2 \, D_o$. If $D_o/L > 5$, i.e., $L < 0.2 \, D_o$ Eq. 8.3.3 can be used for determining the safe pressure.

8.4 CIRCUMFERENTIAL STIFFENERS

Circumferential stiffeners are used in external pressure vessels to improve the rigidity against collapsing. For that purpose the stiffeners themselves should be rigid enough. The value of moment of inertia is the measure for such rigidity. Used for vessels under external pressure each stiffening ring is considered to resist the external load for a distance $L/2$ on either side of the ring (where L is the spacing between rings).

Thus the load per unit circumferential length on the ring at collapse is equal to $L(p_c)$. In Eq. 8.2.2, L was taken unity. Therefore,

$$P = p_c(L) = \frac{24\,E\,I}{D_o^3} \qquad \ldots (8.4.1)$$

Where, P = load on combined shell and stiffener per unit circumferential length.

From Eq. 8.4.1,

$$I = \frac{p_c\,D_o^3\,L}{24\,E}$$

$$= \left(\frac{D_o^2\,L\,t}{12\,E}\right)\left(\frac{p_c\,D_o}{2\,t}\right)$$

$$= \left(\frac{D_o^2\,L\,t}{12\,E}\right)(f_c) \qquad \ldots (8.4.2)$$

As the collapsing pressure is less than theoretically calculated critical pressure p_c, the expression for required moment of inertia will be

$$I = \left(\frac{D_o^2\,L\,t}{12\,E}\right)(f) \qquad \ldots (8.4.3)$$

Where, f = allowable compressive stress,
t = corroded shell thickness,
L = distance between stiffeners,
E = modulus of elasticity at design temperature,
D_o = outer diameter of the shell,
I = moment of inertia.

The moments of inertia of the stiffening ring and the shell act together to resist collapse of the vessel under external pressure. Timoshenko[2] has shown that the combined moment of inertia of the shell and stiffener can be considered as equivalent to that of a thicker shell,

i.e., $\quad t_e = t + \dfrac{A_s}{L} \qquad \ldots (8.4.4)$

where, t_e = equivalent thickness of shell,
A_s = cross-sectional area of one circumferential stiffener.
$\geqslant 3\,t^2$

substituting in Eq. 8.4.3,

$$I = \frac{D_o^2\,L\left(t + \dfrac{A_s}{L}\right)f}{12\,E} \qquad \ldots (8.4.5)$$

where, I = required moment of inertia of stiffening ring.
$\geqslant 2\,t^4$

According to the literature[1] and standard codes, the combined moment of inertia of the stiffener and the shell together is greater than the moment of inertia of the stiffener alone. As a result the Eq. 8.4.5 gives conservative value with additional safety in design.

134 CHEMICAL EQUIPMENT DESIGN—MECHANICAL ASPECTS

Any external metal welded or rigidly held along the circumference can be considered as stiffener provided it satisfies Eq 8.4.5. In determining end side effective length, 1/3 depth (inside) of formed end is to be added to the cylindrical length. Fig 8.1 shows stiffener cross-section and effective length. Moment of inertia of various cross-sections are given in Appendix C.

Design Example 8.1 : A fractionating tower is 4 m in outside diameter by 6 m in length from tangent line to tangent line of closures. The tower contains removable trays on a 1 m tray spacing and is to operate under vacuum at 400 °C. The material of construction is IS : 2002—1962 Gr. I plain carbon steel.

(a) Determine the required thickness of the shell without stiffeners and then with stiffeners located at the tray positions.

(b) Design the stiffening rings. Evaluate if the use of stiffening rings will be advantageous.

Solution : (a) Determination of shell thickness without stiffening rings :

Shell will be designed to work under full vacuum, i.e., $p = 0.1$ MN/m². Design is to be checked both for elastic instability and plastic deformation. Assuming standard dished head (Torispherical) at both ends of the tower having $R_i = D_o$ and $r_i = 0.1 D_o$, where, R_i is the inner crown radius, r_i inner knuckle radius and D_o the outer diameter of the vessel. The inside depth h_i of the end can be calculated from the following correlation (Ch. 4).

$$h_i = R_i - \sqrt{\left(R_i - \frac{D_i}{2}\right) \times \left(R_i + \frac{D_i}{2} - 2 r_i\right)}$$

Where, $R_i = 4$
$D_i = D_o$ (initial approximation) $= 4$
$r_i = 0.4$

Substituting,
$$h_i = 4 - \sqrt{(4 - 2)(4 + 2 - 0.8)}$$
$$= 0.78 \text{ m}$$

Effective length of the tower without stiffener :
$L =$ tangent to tangent length $+ 2 \times (1/3) h_i$
$= 6 + 0.52$
$= 6.52$ m

Therefore,
$$\frac{D_o}{L} = \frac{4}{6.52} = 0.615$$

Modulus of elasticity E for carbon steel at 400 °C is given as :
$$E = 1.67 \times 10^5 \text{ MN/m}^2$$

DESIGN OF PROCESS VESSELS AND PIPES UNDER EXTERNAL PRESSURE 135

Corroded shell thickness required for elastic stability can be obtained from Eq. 8.2.6 and Table 8.2.

$$p = KE\left(\frac{t}{D_o}\right)^m$$

Where, $p = 0.1$
$E = 1.67 \times 10^5$
$K = 0.52$
$m = 2.49$
$D_o = 4$

Substituting,

$$0.1 = 0.52 \times 1.67 \times 10^5 \left(\frac{t}{4}\right)^{2.49}$$

and $t = 16.676 \times 10^{-3}$ m

Checking for plastic deformation :

This is done by using Eq. 8.3.4, as D_o/L is less than 5.

$$p = 2f\left(\frac{t}{D_o}\right) \frac{1}{1 + \dfrac{1.5\, U\, (1 - 0.2\, D_o/L)}{100\left(\dfrac{t}{D_o}\right)}}$$

f = allowable compressive stress,
$\geqslant$ allowable tensile stress
= 70 MN/m²

$\dfrac{t}{D_o} = 4.169 \times 10^{-3}$ (from earlier result).

$U = 1.5\%$ (assumed for new vessel)

$D_o/L = 0.615$

Substituting,

$$p = 2 \times 70 \times 4.169 \times 10^{-3} \times \frac{1}{1 + \dfrac{1.5 \times 1.5\, (1 - 0.2 \times 0.615)}{100 \times 4.169 \times 10^{-3}}}$$

$= 0.101 > 0.1$

Therefore, the calculated thickness is safe against plastic deformation also :

$t = 18$ mm (standard thickness selected).

If stiffeners are used, the effective length of the tower will be reduced to 1 m (1 spacing), i.e.,

$L = 1$

$\dfrac{D_o}{L} = 4$

From Table 8.2, $K = 9.037$ and $m = 2.62$

Substituting in Eq. 8.2.6

$$0.1 = 9.037 \times 1.67 \times 10^5 \, (t/D_o)^{2.62}$$

or, $t/D_o = 1.82 \times 10^{-3}$

or $t = 7.28 \times 10^{-3}$ m

Check for plastic deformation (Eq. 8.3.4)

$$p = 2 \times 70 \times 1.82 \times 10^{-3} \times \frac{1}{1 + \frac{1.5 \times 1.5 \, (1 - 0.2 \times 4)}{0.18}}$$

$$= 0.074 < 0.1$$

The calculated thickness is not sufficient for plastic failure. A new value of t (shell thickness) is to be found out from Eq. 8.3.4.

$$0.1 = 2 \times 70 \times \frac{t}{D_o} \times \frac{1}{1 + \frac{1.5 \times 1.5 \, (1 - 0.2 \times 4)}{100 \left(\frac{t}{D_o}\right)}}$$

$$= 140 \left(\frac{t}{D_o}\right) \times \frac{1}{1 + \frac{4.5 \times 10^{-3}}{(t/D_o)}}$$

Solution of this involves trial and error. To get the exact value, this can be expressed in quadratic form and the result obtained gives :

$$\frac{t}{D_o} = 2.2 \times 10^{-3}$$

or $t = 8.8 \times 10^{-3}$ m

A standard plate thickness of $t = 10$ mm is selected.

(b) Design of stiffening ring involves first to select a standard structure and then to check for required moment of inertia (Eq. 8 4.5) with the moment of inertia of the structure.

From Eq. 8.4.5

$$I = \frac{D_o^2 \, L \left(t + \frac{A_s}{L}\right) f}{12 \, E}$$

where, I = required moment of inertia
D_o = outer diameter of shell = 4 m
L = effective length of tower = 1 m
t = shell thickness = 1.0×10^{-2} m
A_s = cross-sectional area of a stiffening ring
f = allowable compressive stress = 70 MN/m^2

DESIGN OF PROCESS VESSELS AND PIPES UNDER EXTERNAL PRESSURE

E = modulus of elasticity
  = 1.67×10^5 MN/m².

Substituting,

$$I = \frac{(4)^2 (1)\left(0.01 + \dfrac{A_s}{1}\right) \times 70}{12 \times 1.67 \times 10^{-5}}, \text{ m}^4$$

$$= 5.6 \times 10^{-6} (1 + 100 A_s), \text{ m}^4$$

Easy solution for this will be to select first a structural shape, such as, I-section, angle channel, rectangular bar, etc., and then from tables given in Appendix C, select A_s find I from above equation and compare with the actual I-value of the selected cross-section. Required I-value should be less than the actual I of the cross-section.

Select a 18-cm channel of following specification :
Weight = 14.6 kg/m
$A_s = 1.84 \times 10^{-3}$ m²
$I = 8.9 \times 10^{-6}$ m⁴
Required $I = 5.6 \times 10^{-6} (1 + 0.184)$
  $= 6.65 \times 10^{-6}$ m⁴

This is found to be adequate.

No. of stiffeners required = 5

Total weight of rings = $5 \times 3.14 \times 4 \times 14.6$ kg
  = 915 kg

Saving in shell material for using stiffening rings
  = $3.14 \times 4 (18 - 10) \times 10^{-3} \times 6 \times 7\,850$ kg
  = 4 750 kg

Therefore, use of stiffening rings will be advantageous in this case.

8.5 SPHERICAL SHELL UNDER EXTERNAL PRESSURE

The spherical shell under external pressure is subjected to failure by elastic instability. Following equations give the thickness required or safe pressure under which the vessel can operate. The equations are suggested in IS : 2825 – 1969.

$$t = \frac{p\,D_o}{0.8\,\sigma} \qquad \ldots (8.5.1)$$

$$p = \frac{0.8\,\sigma\,t}{D_o} \qquad \ldots (8.5.2)$$

Where, σ = 0.2 per cent proof stress.

8.6 PIPES AND TUBES UNDER EXTERNAL PRESSURE

The relationships presented earlier for long, thin cylinders under external pressure are applicable for $D_o \leqslant 0.2$ m and $D_o/D_i \not> 1.2$. Tubing and pipes usually have $D_o < 0.2$ m and D_o/D_i larger. As a result failure is expected by elastic-plastic buckling rather than by elastic failure. Following equations apply both for external and internal pressures, if D_o/D_i does not exceed 1.7. These equations are suggested in German code (AD Merkblaetter).

$$t_i = \frac{p_i D_o}{2fJ} + c \qquad \ldots (8.6.1)$$

$$t_e = \frac{0.6 p_e D_o}{fJ} + c \qquad \ldots (8.6.2)$$

where,
- t_i = wall thickness under internal pressure,
- t_e = wall thickness under external pressure,
- p_i = design pressure (internal),
- p_e = design pressure (external),
- D_o = outer diameter,
- f = allowable stress at design temperature,
- J = longitudinal joint efficiency,
  - = 0.7 if welded from one side only,
  - = 0.8 if welded from both the sides,
  - = 0.9 if welded from both sides and then annealed.
- c = Corrosion, abrasion allowances.

The minimum wall thickness for seamless and longitudinally welded pipes under external pressure should be as follows:

Table 8.3 Minimum Wall Thickness for Pipes under External Pressure

D_o (mm)		t_e (mm)	D_o (mm)		t_e (mm)
Up to	28	2.00	From	108	3.75
Above	28	2.50	From	121	4.00
From	57	2.75	From	146	4.25
From	60	3.00	From	159	4.50
From	83	3.25	From	191	5.25
From	95	3.50			

For smaller wall thickness and larger external pressure, failure due to elastic deformation is to be checked as per the equations given for cylindrical shell design under external pressure.

Table 8.4 gives the dimensions of black steel tubes.

For tubes and pipes fitted with appropriate flanges or suitably butt welded together, the maximum permissible pressure can be 2.1 MN/m² and maximum temperature 260 °C.

Tubes are available in random lengths from 4 to 7 m.

Table 8.4 Dimensions of Black Steel Tubes*

Nominal Bore (mm)	LIGHT Out side Diam. Max. (mm)	LIGHT Out side Diam. Min. (mm)	LIGHT Thickness (mm)	MEDIUM Out side Diam Max. (mm)	MEDIUM Out side Diam Min. (mm)	MEDIUM Thickness (mm)	HEAVY Out side Diam. Max. (mm)	HEAVY Out side Diam. Min. (mm)	HEAVY Thickness (mm)
6	10.1	9.7	1.8	10.6	9.8	2.00	10.6	9.8	2.65
8	13.6	13.2	1.8	14.0	13.2	2.35	14.0	13.2	2.90
10	17.1	16.7	1.8	17.5	16.7	2.35	17.5	16.7	2.90
15	21.4	21.0	2.00	21.8	21.0	2.65	21.8	21.0	3.25
20	25.9	26.4	2.35	27.3	26.5	2.65	27.3	26.5	3.25
25	33.8	33.2	2.65	34.2	33.3	3.25	34.2	33.3	4.05
32	42.5	41.9	2.65	42.9	42.0	3.25	42.9	42.0	4.05
40	48.4	47.8	2.90	48.8	47.9	3.25	48.8	47.9	4.05
50	60.2	59.6	2.90	60.8	59.7	3.65	60.8	59.7	4.50
65	76.0	75.2	3.25	76.6	75.3	3.65	76.6	75.3	4.50
80	88.7	87.9	3.25	89.5	88.0	4.05	89.5	88.0	4.85
100	113.9	113.0	3.65	115.0	113.1	4.50	115.0	113.1	5.40
120	—	—	—	140.8	138.5	4.85	140.0	138.5	5.40
150	—	—	—	166.5	163.9	4.85	166.5	163.9	5.40

*In accordance with ISO/R 65

CHAPTER 9
DESIGN OF TALL VESSELS

9.1 INTRODUCTION

Self-supporting tall equipment are widely employed in chemical process industries. A few of them are distillation column, absorption tower, multistage reactor, etc. Due to their large height, these equipment are often erected in the open space, rendering them to wind action. As a result, the prediction of membrane stresses due to internal or external pressure will not be sufficient to design such vessels. Special consideration is necessary to take into account to predict the stresses induced due to dead weight, action of wind and seismic forces.

9.1.1 Stresses in the shell

The essential stresses which are exerted in the wall of a tall vertical vessel are :

(a) The circumferential, axial and radial stresses due to pressure or vacuum in the vessel.

(b) The compressive stresses caused by dead loads like self-weight of the vessel including insulation and attached equipment and weight of the contents.

(c) Tensile and compressive stresses due to bending moment caused by wind load acting on the vessel and its attachments.

(d) Stresses due to eccentricity as a result of irregular load distribution for piping, platform, etc.

(e) Shearing stress caused by a torque about the longitudinal axis from offset piping and wind loads.

Besides, sometimes there are stresses resulting from seismic forces, and always there are some residual stresses due to fabrication procedures like cold forming, welding, etc.

It is to be noted that the design based on combined stresses must satisfy the requirement prescribed in Ch. 3 (for internal pressure) and Ch. 8 (for external pressure).

9.2 DETERMINATION OF EQUIVALENT STRESS UNDER COMBINED LOADINGS

As the tall vessels are subjected to loadings additional to those internal or external pressure, the stress equivalent to the membrane stresses should nowhere exceed the allowable stress value of the material of construction. There are four theories reported in the literature[1] to determine the equivalent stress under combined loadings.

---

[1] L. E. Brownwell and E. H. Young "Process Equipment Design", John Wiley and Sons, Inc., New York.

These are:
1. Maximum-principal-stress theory
2. Maximum-shear-energy theory
3. Maximum-strain-energy theory
4. Modified-shear-strain-energy theory.

IS : 2825—1969 suggests to check the stress equivalent to membrane stress determined o the basis of two dimensional maximum-stress theory and two dimensional modified-strai energy theory.

The equivalent stress applying maximum stress theory gives the followin relationship :

$$\sigma_e = \sigma_z \qquad \ldots (9.2.1)$$

The stress equivalent to membrane stress on the shear-strain-energy criterion is given by :

$$\sigma_e = (\sigma_\theta^2 - \sigma_\theta \sigma_z + \sigma_z^2 + 3\tau^2)^{\frac{1}{2}} \qquad \ldots (9.2.2)$$

where, σ_z = resultant longitudinal stress comprising axial tensile or compressive stress due to pressure

$$\left(\text{i.e., } \frac{p D^2}{4 t (D_i + t)} \right)$$

dead load $\left(\text{i.e., } \dfrac{W}{\pi t (D_i + t)} \right)$ and

longitudinal bending moment $\left(\text{i.e., } \dfrac{4 M}{\pi t (D_i + t) D_i} \right)$,

σ_θ = hoop stress

$$= \frac{p (D_i + t)}{2 t} \qquad \ldots (9.2.3)$$

τ = shearing stress

$$= \frac{2 \tau}{\pi t (D_i + t) D_i} \qquad \ldots (9.2.4)$$

p = internal or external pressure (design or test condition)

t = shell thickness (before adding corrosion allowance),

D_i = internal diameter of shell,

M = longitudinal bending moment due to wind action, seismic forces and eccentric loads,

τ = torque about vessel axis.

DESIGN OF TALL VESSELS

W = dead load

(a) for points above plane of support :
weight of shell, closure, insulation, fittings, attachments and fluid supported above the point considered, causes compressive stress.

(b) for points below plane of support :
weight of vessel, fittings and attachments below the point considered, plus weight of fluid contents, causes tensile stress in the shell below the point considered.

For the safe design following requirements are to be fulfilled at design and test conditions (IS : 2825—1969).

At design conditions :

$\sigma_\theta \leqslant f \cdot J$ ... (9.2.5a)
σ_z (tensile) $\leqslant f \cdot J$ ... (9.2.5b)
σ_z (compressive) $\leqslant 0.125 \, E \, (t/D_o)$ ... (9.2.5c)

At test conditions :

$\sigma_\theta \leqslant 1.3 \, f_a \, J$ ... (9.2.5d)
σ_z (tensile) $\leqslant 1.3 \, f_a \, J$ ... (9.2.5e)
σ_z (compressive) $\leqslant 0.125 \, E_a \, (t_a/D_o)$ ... (9.2.5f)

Where, f = allowable stress at design temperature,
f_a = allowable stress at test temperature,
t_a = actual shell thickness at the time of test,
E = young modulus at design temperature,
E_a = young modulus at test temperature,
D_o = outside diameter of shell,
J = Weld-joint efficiency factor, if circumferential weld is falling at the point under consideration.

Test pressure is determined as follows (IS : 2825—1969) :

Test pressure = $1.3 \times$ design pressure $\times \dfrac{f_a}{f}$

The loads acting on a tall vessel will depend upon the conditions under which the vessel exists.

There may be the following possible cases :

Case 1. Vessel under construction
(a) empty shell erected
(b) shell and auxiliary equipment such as trays or packing but no insulation

Case 2. Vessel completed but not under operation

Case 3. Vessel under test conditions
   (a) hydraulic test
   (b) pneumatic test

Case 4. Vessel in operation

While designing, the possibility of failure under any of the above cases should be checked.

In the consideration of wind and earthquake loads it is assumed that the possibility of occurring simultaneously the most adverse wind and earthquake load will be remote. Therefore, in computing resulting longitudinal stress, either wind load or earthquake load, whichever is most adverse is to be considered. If both the loads are considered together, it may give overdesign.

Design calculations for the thickness should be started from the top of the vessel. The thickness of the upper course of the vessel is usually controlled by the circumferential stress resulting from internal or external pressure. At the lower sections of the vessel, the wall thickness is to be calculated taking into account the combined stresses resulting from increased weight, wind load, etc.

9.3 DETERMINATION OF LONGITUDINAL STRESSES

9.3.1 The axial stress (tensile and compressive) due to pressure is :

$$\sigma_{zp} = \frac{p\, D^2}{4\, t\, (D_i + t)} \qquad \ldots (9.3.1)$$

Where, $D = D_i$ for internal pressure,
   $= D_o$ for vacuum (D_o should include insulation thickness if insulated)

9.3.2. The axial stress (compressive) due to dead loads

9.3.2.1 *The stress induced by shell weight at a distance X meter from the top is :*

$$\sigma_{zs} = \frac{W_s}{\pi\, t\, (D_i + t)} \qquad \ldots (9.3.2)$$

Where, W_s = weight of the shell for a length X meter,
   t = shell thickness at the point under consideration.

9.3.2.2 *The stress induced in the shell due to insulation at a distance X meter from the top is :*

$$\sigma_{zi} = \frac{W_i}{\pi\, t\, (D_i + t)} \qquad \ldots (9.3.3)$$

Where, W_i = weight of insulation upto a distance X meter from the top,
   t = shell thickness at the point under consideration,
   D_i = inside diameter of shell.

DESIGN OF TALL VESSELS

9.3.2.3 *The stress induced by the weight of the liquid supported by the inner arrangement like tray for a distance X meter from the top is*

$$\sigma_{zl} = \frac{W_l}{\pi t (D_i + t)} \quad \ldots (9.3.4)$$

Where, W_l = wt of the liquid supported for a distance X meter from the top.

9.3.2.4 *The axial stress due to the weight of the attachments like trays, overhead condensers, top head, platforms and ladders for a distance X meter from the top is*

$$\sigma_{za} = \frac{W_a}{\pi t (D_i + t)} \quad \ldots (9.3.5)$$

Where, W_a = weight of all attachments upto a distance X meters from the top.

As available in the literature[1,2] the weight of steel platforms may be estimated at 1.67 kN/m² of area, and weight of steel ladders at 365 N/m linear length of caged ladders and 146 N/m linear length for plain ladders. Trays in distilling columns, including liquid hold-up on trays, may be estimated to have a weight of 1.2 kN/m² of tray area.

The total dead load stress, σ_{zw}, acting along the axial direction of shell at point is then given by :

$$\sigma_{zw} = \sigma_{zs} + \sigma_{zi} + \sigma_{zl} + \sigma_{za} \quad \ldots (9.3.6)$$

If the vessel does not contain internal attachments like trays which support liquid, the consists only of the shell insulation, the heads, and minor attachments like manholes, nozzles, etc., the additional load may be estimated as approximately equal to 18% of the weight of a steel shell.[1]

9.3.3 The longitudinal bending stresses due to dynamic loads

9.3.3.1 *The axial stresses (tensile and compressive) due to wind loads in self-supporting tall vessels*

The stress due to wind load may be calculated by treating the vessel as uniformly loaded cantilever. The wind load is a function of the wind velocity, air density, shape of the tower, and the arrangement of all such tall vessels.

The wind load on a vessel is given by :

$$P_w = \frac{1}{2} C_D \rho V_w^2 A \quad \ldots (9.3.7)$$

Where, C_D = drag coefficient,
ρ = density of air,
V_w = wind velocity,
A = projected area normal to the direction of the wind.

---
[a] J. G. Nelson, "Calculation sheets for vessel design", Petroleum Refiner, 27, No. 11 (1948), p. 365.

The empirical equivalent of Eq. 9.3.7 has been given by several authors[1]. If wind velocity is known, approximate wind pressure can be computed from the following simplified relationship.

$$p_w = 0.05 \, V_w^2 \qquad \ldots (9.3.8)$$

where, p_w = minimum wind pressure to be used for moment calculation, N/m^2,

V_w = maximum wind velocity experienced by the region under worst weather condition, km/h.

It is also known that the wind velocity changes with the height. The velocity of wind near the ground is less than that away from it. To take into account this factor a variable wind force may be taken.

Since the wind pressure does not remain constant throughout the height of the tall vessel, it is recommended to calculate the wind load in two parts[3]. In case of vessel with height 20 m, it is suggested that the wind load may be determined separately for the bottom part of the vessel having height equal to 20 m, and then for rest of the upper part (Fig. 9.1).

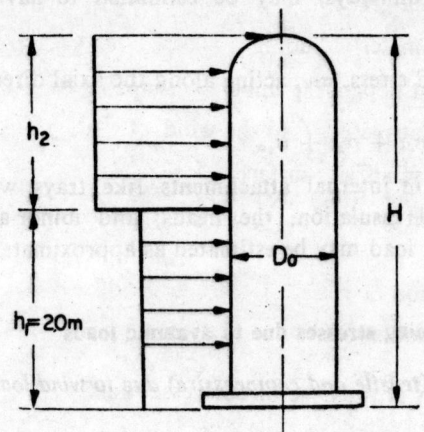

Fig. 9.1 Tall vessel under wind pressure

The wind pressure for bottom part and rest of upper part can directly be obtained from the Table 9.1 depending upon the zone where the vessel will be installed.

Table 9.1 Wind Pressure[3]

Nature of the region	Wind pressure (kN/m²)	
	at H = 20 m	at H = 100 m
Coastal area	0.7 – 1.0	1.5 – 2.0
Area with moderate wind	0.4	1.0

[3] N. I. Taganov, "Fundamentals of Process Equipment Design", Academic Books, Delhi.

The total load due to wind acting on the bottom and upper parts of the vessel are determined from the following equations.

$$P_{bw} = K_1 K_2 p_1 h_1 D_o \qquad \ldots(9.3.9)$$
$$P_{uw} = K_1 K_2 p_2 h_2 D_o \qquad \ldots(9.3.10)$$

Where, P_{bw} = total force due to wind load acting on the bottom part of the vessel with height equal to or less than 20 m,

P_{uw} = total force due to wind load acting on the upper part above 20 m,

h_1 = height of the bottom part of the vessel equal to or less than 20 m,

h_2 = height of the upper part above 20 m,

p_1 = wind pressure for bottom part of the vessel (from Table 9.1, value given for $H = 20$ m),

p_2 = wind pressure for upper part of the vessel (to be determined from Table 9.1 for the mid point of upper part of the vessel by interpolation of the data given),

D_o = outer diameter including insulation as the case may be,

K_1 = coefficient depending upon the shape factor.

= 1.4 for flat plate 90° to the wind

= 0.7 for cylindrical surface

K_2 = coefficient depending upon the period of one cycle of vibration of the vessel

= 1 (if period of vibration is 0.5 seconds or less)

= 2 (if period exceeds 0.5 seconds)

The bending moment at the base of the vessel due to wind load is determined from the following equations.

(i) For the vessels with $H \leqslant 20$ m

$$M_w = P_{bw} \frac{H}{2} \qquad \ldots(9.3.11)$$

(ii) For the vessels with $H > 20$ m

$$M_w = P_{bw} \frac{h_1}{2} + P_{uw} \left(h_1 + \frac{h_2}{2} \right) \qquad \ldots(9.3.12)$$

The resulting bending stress in the axial direction is then computed from the following correlation.

$$\sigma_{zwm} = \frac{4 M_w}{\pi t (D_i + t) D_i} \qquad \ldots(9.3.13)$$

Where, σ_{zwm} = longitudinal stress due to wind moment (compressive on down-wind side and tensile on upwind side),

M_w = bending moment due to wind load,

D_i = inner diameter of the shell,

t = corroded shell thickness.

9.3.3.2 *The stresses due to seismic loads (compressive and tensile occurring simultaneously)*

The seismic forces act to produce horizontal shear in self-supporting vertical vessels. This shear force in turn produces a bending moment about the base of the vessel. The shear loading is triangular with the apex at the base as shown in Fig. 9.2.

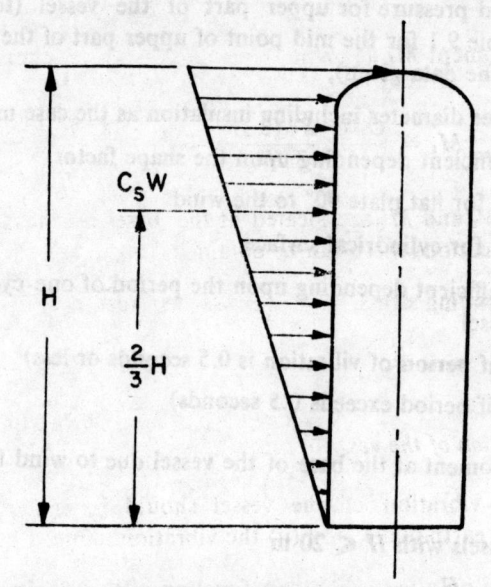

Fig. 9.2 Seismic forces on a tall vessel

If the acceleration of the earthquake is 'a', the force on the vessel will be,

$$F = a \left(\frac{W}{g} \right) = C_s W \qquad \ldots(9.3.14)$$

Where, W = total weight of the vessel,

$C_s = a/g$ is termed as seismic coefficient

The seismic coefficients, C_s, as a function of period of vibration, T, in seconds, are given in Table 9.2.

Table 9.2 Seismic Coefficient[1]

Seismic zone	Seismic coefficient C_s		
	$T < 0.4$	$0.4 < T < 1$	$T > 1.0$
Mild	0.05	$0.02/T$	0.02
Medium	0.10	$0.04/T$	0.04
Severe	0.20	$0.08/T$	0.08

The total shear load resulting from seismic forces is $C_s W$ (Eq. 9.3.14) and its centre of action for such triangular loading is located at $\tfrac{2}{3} H$. The shear load, V_s at any horizontal plane in the tower X meters down from the top is given by[4]:

$$V_s = \frac{C_s W X (2H - X)}{H^2} \qquad \ldots(9.3.15)$$

The bending moment M_s at plane X resulting from the shear forces above plane X is given by:

$$M_s = \frac{C_s W X^2 (3H - X)}{3 H^2} \qquad \ldots(9.3.16)$$

The maximum V_s and M_s are located at the base of the tower and substituting $X = H$ in the above equations these can be obtained.

The resulting bending stress, $\sigma_{z,sm}$, from seismic load is given by:

$$\sigma_{z,sm} = \frac{4 M_s}{\pi t (D_i + t) D_i} \qquad \ldots(9.3.17)$$

9 3.3.3 Period of vibration of the vessel

The study of the vibration of the vessel should be made to understand (i) the vibration induced by the earthquake, and (ii) the vibration induced by the wind.

The seismic force on a vessel is also a function of the flexibility of the system and therefore it becomes necessary to determine the natural frequency of the vessel in order to calculate the magnitude of the earthquake force.

The wind induced vibrations may some times become very severe. When the wind reaches a particular wind velocity, it is interesting to observe, the vessel starts vibrating severely in the direction right angles to the direction of wind due to vortex shedding[4]. These transverse oscillations are caused by the alternate shedding of vortices from the airflow at two sides of the vessel. At the critical wind velocity the frequency of vortex shedding coincides with the natural frequency of the vessel and resonant oscillation begins.

---

[4] "Proceedings of Advanced Summer School on Analysis and Design of Pressure Vessels and Piping", M.N.R.E.C., Allahabad, May, 1972.

The frequency of vortex shedding, f, is given by:

$$f = \frac{S V_w}{D_o} \qquad \ldots(9.3.18)$$

where, S = characteristic number
  = 0.2 for circular cylinders
 V_w = wind velocity,
 D_o = outside diameter of vessel including insulation, if any,

If the natural frequency, N, of the vessel is equated to vortex shedding frequency, f, the critical wind velocity for the vessel will be obtained as follows.

$$N = \frac{S V_c}{D_o} \qquad \ldots(9.3.19)$$

or, $V_c = 5 N D_o \qquad \ldots(9.3.20)$

where, V_c = critical wind velocity.

To avoid the resonant oscillation the vessel should be designed, as far as possible, to avoid critical wind velocity. However, some times it may not be possible to avoid this eventuality and then efforts should be made to either change the natural frequency of the vessel or to change the pattern of vortex shedding. This is achieved by distributing the external attachments (ladders, platforms, piping, etc.) all around the vessel or providing helical strakes fitted to the top of the vessel to break up the continuity of the airflow pattern, or putting a section on the top of vessel filled with water and similar other methods.

During earthquake the sudden acceleration in earth's crust starts vibration in the vessel, as the inertia of the vessel restrains it from moving simultaneously with the foundation. The vibration of the vessel is same as the vibration of a cantilever. The relationship for natural frequency of such a system is derived by considering the motion of a weight suspended on the end of a completely elastic spring[1]. If T is the period of vibration, then

$$N = 1/T \qquad \ldots(9.3.21)$$

The expression for T is given by[1]:

$$T = \frac{2\pi}{3.53} \sqrt{\frac{W H^3}{E I g}} \qquad \ldots(9.3.22)$$

Where, T = period of vibration,
 W = total weight of the tower,
 H = total tower height including skirt,
 E = modulus of elasticity,

DESIGN OF TALL VESSELS

I = moment of inertia of the shell cross-section,

$$= \frac{\pi D^3}{8} t$$

$D = D_i + t$

g = gravitational constants,

Units in Eq. 9.3.22 must be consistant.

If W = total weight in kN,
    $g = 9.81$ m/s^2,
    $E = 2 \times 10^{11}$ N/m^2,

Eq. 9.3.22 can be expressed in the following form:

$$T = 6.35 \times 10^{-5} (H/D)^{3/2} (W/t)^{1/2}, \text{ s.} \qquad \ldots(9.3.23)$$

Where, t = corroded wall thickness, m.

The vessel having a period less than 0.4 second may be regarded as rigid and therefore a high seismic coefficient should be assumed (Table 9.2). If T is greater than one second, the vessel is flexible and the seismic coefficient will be low as slender columns are more able to absorb the seismic forces. On the other hand, the reverse is true under the influence of wind forces. If the column is rigid, it will withstand higher wind forces.

In this connection the influence of stiffeners can be considered. By the addition of internal or external members attached to the shell the rigidity of vessels increases. These attached members may be longitudinal or circumferential or both. Some of the attachments like tray support rings provide stiffening to the tower against buckling. If the stiffening effect is available from existing structure, it is advantageous. But it will not be economical, if some structures are to be attached especially for stiffening only. In that case it will be advisable to increase the wall thickness. It is discussed in the previous paragraph that rigid structure withstands higher wind loads, but may fail against seismic forces.

As the addition of stiffeners increases the buckling strength of the vessel against compressive load, in Eqs. 9.2.5 c and 9.2.5 f, the thickness, t or t_a may be substituted by equivalent thickness, t_e, when $t_e = \sqrt{t_x \, t_y}$, and t_x and t_y are the equivalent thickness of the shell in the longitudinal and circumferential direction respectively.

$$t_x = (t \text{ or } t_a) + \frac{A_x}{d_x}$$

$$t_y = (t \text{ or } t_a) + \frac{A_y}{d_y}$$

Where, A_x = cross-sectional area of one longitudinal stiffener,
    A_y = cross-sectional area of one circumferential stiffener,

d_x = spacing between longitudinal stiffeners,

d_y = spacing between circumferential stiffeners.

9.3.4 The longitudinal bending stresses due to eccentric loads

In tall vessels the externally attached equipment and parts usually act as eccentric loads and should be considered as such. Most of the externally attached parts like small ladders, pipes, manholes, etc. produce negligible moment and can be disregarded. But some of the moments produced by eccentrically placed overhead or side condensers large platforms are important. The eccentricity is calculated by :

$$e = \frac{\Sigma M_e}{\Sigma W_e} \qquad \qquad \text{..(9.3.24)}$$

where, e = eccentricity, the distance from the column axis to the centre of resultant reaction,

ΣM_e = summation of moments of eccentric loads,

ΣW_e = summation of all eccentric loads.

Again, $\Sigma M_e = W_{e1} e_1 + W_{e2} e_2 + \ldots\ldots$ ...(9.3.25)

$\Sigma W_e = W_{e1} + W_{e2} + \ldots\ldots$ ...(9.3.26)

The additional bending stress at plane X caused by this moment is :

$$\sigma_{ze} = \frac{4 \Sigma M_e}{\pi t (D_i + t) D_i} \qquad \qquad \text{...(9.3.27)}$$

9.4 DETERMINATION OF RESULTANT LONGITUDINAL STRESS

The resultant longitudinal tensile and compressive stresses are obtained by appropriate combination of various stresses determined in section 9.3.

The resulting tensile stress (on upwind side) in the cross-section of the vessel at a distance X meters from the top in absence of eccentric load will be :

(i) for internal pressure

$$\sigma_z = \sigma_{zp} - \sigma_{zw} + (\sigma_{zwm} \text{ or } \sigma_{zsm}) \qquad \text{...(9.4.1)}$$

(ii) for external pressure

$$\sigma_z = (\sigma_{zwm} \text{ or } \sigma_{zsm}) - \sigma_{zp} - \sigma_{zw} \qquad \text{...(9.4.2)}$$

The resulting compressive stress (on downwind side) in the cross-section of the vessel at a distance X meters from the top in absence of eccentric load will be :

(i) for internal pressure

$$\sigma_z = (\sigma_{zwm} \text{ or } \sigma_{zsm}) + \sigma_{zw} - \sigma_{zp} \qquad \text{...(9.4.3)}$$

(ii) for external pressure

$$\sigma_z = (\sigma_{zwm} \text{ or } \sigma_{zsm}) + \sigma_{zw} + \sigma_{zp} \qquad \text{...(9.4.4)}$$

Equivalent stress, σ_e is obtained from Eq. 9.2.2 substituting σ_z and stresses are checked with the conditions given in Eqs. 9.2.5.

The Equations cannot be reduced to a convenient explicit expression for the calculation of minimum wall thickness, t, without any approximation and therefore the required corroded thickness is to be found out by trial and error. As the minimum wall thickness cannot be less than the value required to satisfy the hoop stress equation the thickness calculated by that can be taken as starting point.

Table 9.3 gives the sp. weight of a few commonly used insulating materials for ready reference.

Table 9.3 Specific Weight of Insulating Materials[5]

Material	Apparent sp. weight kN/m^3	Thermal conductivity $W/m\ °C$
Asbestos	5.64	0.195
Chalk	15.00	0.692
Plaster, artificial	20.70	0.742
Cotton wool	0.78	0.042
Cork board	1.57	0.043
Cork, ground	1.45	0.043
Diatomaceous earth powder, fine	2.70	0.069
Felt, wool	3.26	0.052
Graphite, powdered	4.78	0.180
Gypsum, molded & dry	12.20	0.432
Magnesia, powdered	7.84	0.605
Mineral wool	1.45	0.041
Rubber, hard	11.70	0.150
Sawdust	1.88	0.052
Silk	0.99	0.045
Wool, animal	1.08	0.036

[5] D. Q. Kern, "Process Heat Transfer", McGraw-Hill Book Co., Inc., New York.

Design Example 9.1: A process sketch of a functionating tower is given in Fig. 9.3. The vessel has the following specifications. Shell thickness is to be determined. As a general guide it is to be noted that for each 5-6 m height, shell thickness can be increased by 1-2 mm. This is required to determine the number of shell courses.

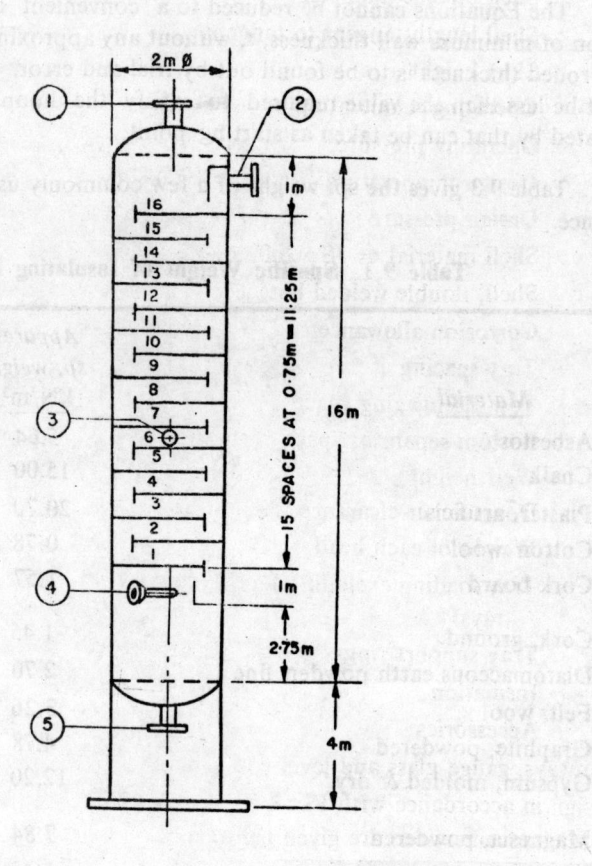

Fig. 9.3 Process design sketch of distillation column

DESIGN OF TALL VESSELS

Max wind velocity expected (for height upto 20 m) = 140 km/h
Shell outside diameter = 2.0 m
Shell length tangent to tangent = 16.0 m
Skirt height = 4.0 m
Operating temperature = 300 °C
Operating pressure = 0.7 MN/m²
Design temperature = 320 °C
Design pressure = 0.8 MN/m²
Shell material = IS : 2002 — 1962 Grade 2 B
Shell, double welded butt joints, no stress relieving or radiographing.
Corrosion allowance = 3 mm
Tray spacing = 0.75 m
Top disengaging space = 1.0 m
Bottom separator space = 2.75 m
Weir height = 75 mm all trays
Downcomer clearance = 25 mm all trays
Weight of each head = 7.5 kN
Tray loading excluding liquid (alloy steel trays) = 1.0 kN/m² of tray area
Tray support rings = 60 mm × 60 mm × 10 mm angles
Insulation = 75 mm asbestos
Accessories = one caged ladder

Manways, gauge glass and level control connections are provided.
Design in accordance with IS : UPV code for class 2 vessels neglecting seismic forces.
Nozzles specifications are given below :

No.	Size (m)	Service
1	0.400	Overhead vapour
2	0.075	Reflux
3	0.300	Feed inlet, internal distributor designed to provide even distribution across length of tray
4	0.400	Reboiler, made tangential
5	0.200	Bottoms.

Solution : Allowable stress of shell material at design temperature = 98.1 MN/m²
Weld-joint efficiency factor = 0.85
Both the values are obtained from IS : 2825 — 1969
Thickness of shell required for internal pressure :

$$t_s = \frac{p\, D_o}{2fJ + p} + c$$

Substituting for

$$p = 0.8 \text{ MN/m}^2$$
$$D_o = 2.0 \text{ m}$$
$$f = 98.1 \text{ MN/m}^2$$
$$J = 0.85$$
$$c = 0.003 \text{ m}$$
$$t_s = 0.012\ 6 \text{ m}$$

Nearest standard thickness = 14 mm
$$= 0.014 \text{ m}$$

Corroded shell thickness, $t = 0.011$ m

As the shell thickness is very small compared to the diameter, for rest of the calculations, $D_o = D_i = (D_i + t) = 2.0$ m will be used.

Calculation of axial stress due to pressure :

$$\sigma_{ap} = \frac{p\,D}{4\,t} = \frac{0.8 \times 2}{4 \times 0.011} = 36.3 \text{ MN/m}^2 \quad \text{.. (by Eq. 9.3.1)}$$

Calculation of axial stress due to dead loads :

W_s = wt. of shell for X meters length
$$= (\pi\,D\,t\,X\,\gamma_s),\ N$$

$$\sigma_{zs} = \frac{W_s}{\pi\,t\,D} = (\gamma_s)\,(X) \times 10^{-6} \text{ MN/m}^2$$
$$= (9.81)\,(7\ 850)\,(X) \times 10^{-6} \text{ MN/m}^2$$
$$= 0.077\,X \text{ MN/m}^2 \quad \text{... (by Eq. 9.3.2)}$$

(Assuming constant thickness of shell)

W_i = weight of insulation for a length X meters
$$= (\pi\,D_{ins}\,t_{ins}\,X\,\gamma_{ins}) \times 10^{-6} \text{ MN}$$

D_{ins} = mean diam. of insulation
$$\approx D$$

$$\sigma_{zi} = \frac{W_i}{\pi\,D\,t} \quad \text{... (by Eq. 9.3.3)}$$
$$= \frac{t_{ins}\,\gamma_{ins}\,X}{t}$$
$$= \frac{(0.075)\,(5\ 640)\,(X) \times 10^{-6}}{0.011}$$
$$= 0.038\ 5\,(X) \text{ MN/m}^2$$

W_l = weight of liquid supported for a distance X meters from top

DESIGN OF TALL VESSELS

No. of trays, $\quad n = \left(\dfrac{X-1}{0.75} + 1\right) = \left(\dfrac{4X-1}{3}\right)$

Liquid weight on trays are calculated on the basis of water and 0.075 m depth. Hence

$$W_l = \dfrac{\pi}{4}(D)^2 (0.075)(9\,810)\left(\dfrac{4X-1}{3}\right) \times 10^{-6} \text{ MN}$$

$$\sigma_{zl} = \dfrac{W_l}{\pi D t} \qquad \ldots \text{(by Eq. 9.3.4)}$$

$$= \dfrac{(75)(9.81 \times 10^{-6})\left(\dfrac{4X-1}{3}\right)}{2(0.011)}$$

$$= 0.033\,5\left(\dfrac{4X-1}{3}\right) \text{ MN/m}^2$$

Weight of tophead $= 7.5 \times 10^{-3}$ MN

Weight of ladder $= 3.65 \times 10^{-4}(X)$ MN

Weight of trays $= \left(\dfrac{\pi D^2}{4}\right)(1)\left(\dfrac{4X-1}{3}\right) \times 10^{-3}$ MN

Hence,

$W_a = $ weight of attachments

$$= (7.5 \times 10^{-3}) + (3.65 \times 10^{-1} X) + (3.14 \times 10^{-3})\left(\dfrac{4X-1}{3}\right), \text{ MN}$$

$$\sigma_{za} = \dfrac{W_a}{\pi D t} \qquad \ldots \text{(by Eq. 9.3.5)}$$

$$= 0.11 + 5.3 \times 10^{-3}(X) + 0.046\left(\dfrac{4X-1}{3}\right), \text{ MN/m}^2$$

$$= 0.066\,7(X) - 0.223, \text{ MN/m}^2$$

By Eq. 9.3.6

$$\sigma_{zw} = 0.077 X + 0.038\,5 X + (0.044\,8 X - 0.011\,2) + 0.056\,7 X - 0.223$$

$$= 0.227 X - 0.234$$

Calculation of stress due to wind loads:

By Eq. 9.3.8, wind pressure p_w, is calculated.

$$p_w = 0.05 V_w^2$$
$$= 0.05 (140)^2$$
$$= 980 \text{ N/m}^2$$

From Table 9.1, maximum wind pressure upto 20 m height is 1 000 N/m². For calculating wind load, $p_w = 1.0$ kN/m² will be used.

By Eq. 9.3.9, wind load, P_w, is:
$$P_w = K_1 K_2 p_w X D_o$$
Where, $p_w = 1\,000$ N/m^2
$$D_o = 2.0 + 0.075 \times 2$$
$$= 2.15 \text{ m (here effect of ladder and piping are not considered)}$$
$$K_1 = 0.7$$

To decide for K_2, period of variation, T, is to be known.

From Eq. 9.3.23

$$T = 6.35 \times 10^{-5} \, (H/D)^{\frac{3}{2}} \, (W/t)^{\frac{1}{2}}, \text{ second}$$

Where, $H = 20$ m
$D = 2$ m
$t = 0.011$ m
$$W = W_s + W_i + W_l + W_a$$
$$= \pi Dt (H-4) \gamma_s + \pi D t_{ins} (H-4) \gamma_{ins} +$$
$$\frac{\pi}{4} D^2 (0.075) \gamma_1 (16) + [7.5 \times 2 + 0.365 (H-4) + \frac{\pi}{4} D^2 (1.0)(16)] \cdot$$
$$= (3.14)(2)(16)(0.011)\,77 + (3.14)(2)(0.075)(16)(5.64)$$
$$+ (0.785)(4)(0.075)(9.81)(16) + (15.0) + 5.85 + 50.15)$$
$$= 84.00 + 42.00 + 37.00 + 71.00 = 234 \text{ kN}$$

Therefore,
$$T = 6.35 \times 10^{-5} \, (20/2)^{\frac{3}{2}} \, (234/0.011)^{\frac{1}{2}}$$
$$= 0.30 \text{ s} < 0.5 \text{ s}$$

Hence,
$$K_2 = 1$$
Substituting,
$$P_w = (0.7)(1)(1\,000)\, X \,(2.15), \text{ N}$$
$$= 1\,500\, X, \text{ N}$$

By Eq. 9.3.11
$$M_w = P_w \frac{X}{2}, \text{ J}$$
$$= 750\, X^2, \text{ J}$$

By Eq. 9.3.13
$$\sigma_{zwm} = \frac{4 M_w}{\pi D^2 t} \, (10^{-6}) \text{ MN/m}^2$$

DESIGN OF TALL VESSELS

$$= \frac{(4)(750) X^2 (10^{-6})}{(3.14)(4)(0.011)} \text{ MN/m}^2$$

$$= 0.021\ 7\ X^2 \text{ MN/m}^2$$

Calculation of resultant longitudinal stress:

Upwind side:

By Eq. 9.4.1

σ_z (tensile) $= \sigma_{zp} - \sigma_{zw} + \sigma_{zwm}$

$= 36.3 + 0.234 - 0.227 X + 0.021\ 7\ X^2$

$= 0.021\ 7\ X^2 - 0.227 X + 36.534$

Expecting, circumferential weld-seam may fall at a distance X meters from top or a little earlier, by Eq. 9.2.5b

σ_z (max) $= fJ$

$= 98.1 \times 0.85$

$= 83.0 \text{ MN/m}^2$

Substituting σ_z (max) for σ_z (tensile)

$83 = 0.021\ 7\ X^2 - 0.227 X + 36.534$

or, $0.021\ 7\ X^2 - 0.227 X - 46.47 = 0$

Solving for X,

$$X = \frac{0.227 \pm \sqrt{(0.227)^2 + (4)(0.021\ 7)(46.47)}}{(2)(0.021\ 7)}$$

$= 51.5 \text{ m} >> 16 \text{ m}$

Downwind side:

By Eq. 9.4.3

σ_z (compression) $= \sigma_{zwm} = \sigma_{zw} + \sigma_{zp}$

$= 0.021\ 7\ X^2 + 0.227 X - 0.234 - 36.3$

$= 0.021\ 7\ X^2 + 0.227 X - 36.534$

By Eq. 9.2.5c

σ_z (compressive, maximum) $= 0.125\ E\ (t/D_o)$

$= 0.125\ (2 \times 10^5) \left(\frac{0.011}{2}\right)$

$= 137.5 \text{ MN/m}^2$

Equating the maximum value of σ_z (compressive),

$0.021\ 7\ X^2 + 0.227 X - 174.034 = 0$

or, $X = 83.5 \text{ meters} >> 16 \text{ m}$

If credit for reinforcement of the shell by tray-support rings are also taken into account, X value will be further increased.

From the longitudinal stress consideration alone, it is observed that hoop stress is controlling the design and a uniform thickness of 14 mm throughout the shell length is sufficient.

Now, the design is to be checked from equivalent stress consideration.

By Eq. 9.2.2

$$\sigma_e = (\sigma_\theta^2 - \sigma_\theta \sigma_z + \sigma_z^2)^{\frac{1}{2}}$$

(assuming no shear stress)

By Eq. 9.2.3

$$\sigma_\theta = \frac{p(D_i + t)}{2t} = \frac{p(D_o - t)}{2t}$$

$$= \frac{0.8(2 - 0.011)}{2 \times 0.011} = 72.5 \text{ MN/m}^2$$

According to earlier calculation,

$$\sigma_z \text{ (tensile)} = 0.021\ 7\ X^2 - 0.227\ X + 36.534$$

Substituting, $X = 16$ m i.e. the tower height.

The maximum induced longitudinal stress at the skirt junction of the shell will be obtained.

or, $\sigma_z = 0.021\ 7\ (16)^2 - 0.227\ (16) + 36.534$

$= 5.55 - 3.632 + 36.534$

$= 38.45$ MN/m²

Substituting these values,

$$\sigma_e = [(72.5)^2 - (72.5)(38.45) + (38.45)^2]^{\frac{1}{2}}$$

$= 63.0$ MN/m² $< fJ$ i.e. 83.0 MN/m²

Stress equivalent to membrane stress is also within the limit prescribed by Eq. 9.2.5a.

Therefore, under operating condition, a shell thickness of 14 mm will be safe.

This design can be checked for other conditions as described in section 9.2, considering vessel under construction, vessel completed and vessel under testing.

In this problem hoop stress is controlling and therefore, other case calculations are not shown.

CHAPTER 10

DESIGN OF SUPPORT FOR PROCESS VESSELS

10.1 INTRODUCTION

Design of a process vessel cannot be complete without the selection and design of a suitable support for it, and also without examining the effect of support on shell. A process vessel is usually supported either in vertical or horizontal position depending upon the process requirement. For example, a distillation column, an absorption tower, an evaporator or a stirred tank reactor will always be supported in vertical position. On the other hand a heat exchanger or condenser or a storage vessel can be supported either in vertical or horizontal position depending upon the floor area and head room available.

For vertical vessels the common supports are :

    (i) Skirt support

    (ii) Bracket or lug support

    (iii) Leg support

    (iv) Ring support

For horizontal vessels the common supports are :

    (i) Saddle support

    (ii) Leg support

    (iii) Ring support

Choosing a particular type of horizontal or vertical supports will depend on size, shape and weight of the vessel, on the design temperature and pressure, on the location of various connections and on the arrangement of the internal and external structures. When differential thermal expansion has to be considered, as in boiler drums, the vessel is usually supported from slings or hangers.

10.2 DESIGN OF SKIRT SUPPORT

Skirt supports are cylindrical or conical steel shells attached to the bottom tangent of the vertical vessels. These supports are found to be most suitable for tall vessels subjected to longitudinal bending stresses. The section modulus of skirt is maximum for a given metal sectional area, as it is concentrated at the maximum distance from the axis of the vessel, and therefore, this type of supports is very suitable structure for vessels subjected to wind, seismic and other loadings, which cause bending moment at the base of the vessel. It is to be noted that the skirts are not subjected to operating pressure conditions as in the case of pressure vessels, and therefore, the selection of material is not limited to the steels permitted by the pressure-vessel codes. The structural steels with corresponding allowable stress may be used for the skirt, and this will be economical. Fig. 10.1 shows cylindrical and conical skirt supports.

A. CYLINDRICAL SKIRT   **B. CONICAL SKIRT**

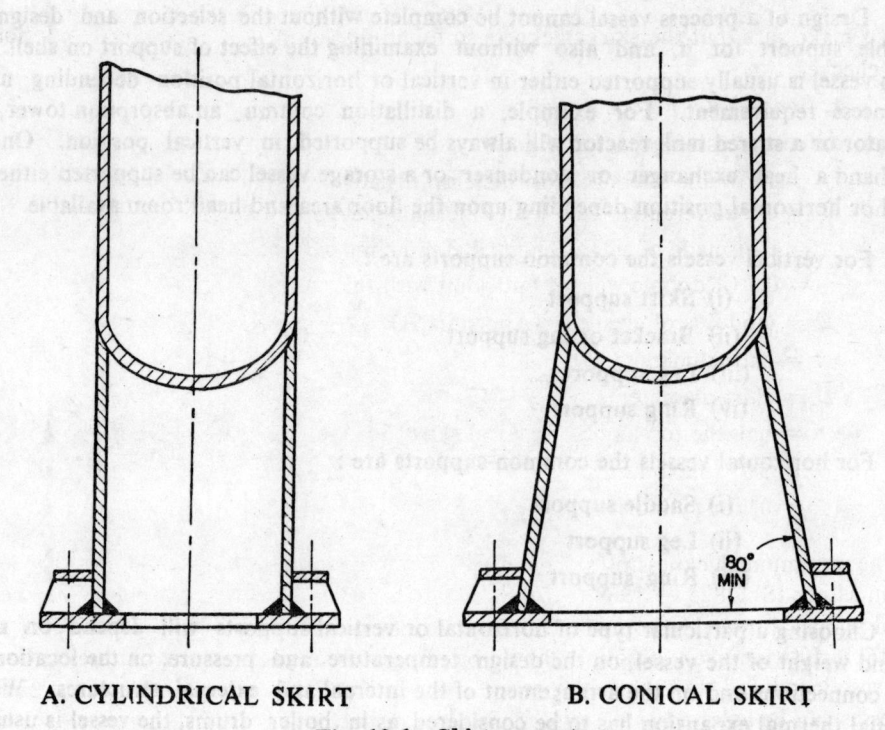

Fig. 10.1 Skirt supports

10.2.1 Skirt-wall thickness

The maximum stress will be induced in the skirt due to the action of the dead weight of the vessel and the wind or seismic bending moment, as the skirt is not subjected to internal or external pressure, like the vessel's shell. The equations, which are derived in Chapter 9 for tall vertical vessels, are applicable for skirt design with a little alteration.

The maximum tensile stress in the skirt wall is given by :

$$\sigma_z \text{ (tensile)} = (\sigma_{zwm} \text{ or } \sigma_{zsm}) - \sigma_{zw} \qquad \ldots (10.2.1)$$

The maximum compressive stress in the skirt wall is obtained from:

$$\sigma_z \text{ (compressive)} = (\sigma_{zwm} \text{ or } \sigma_{zsm}) + \sigma_{zw} \qquad \ldots (10.2.2)$$

Where, σ_{zw} = the stress due to total dead weight supported by the skirt (to be evaluated at the base of the skirt by Eq. 9.3.6),

σ_{zwm} = the stress due to wind moment at the base of the skirt (Eq. 9.3.13),

σ_{zsm} = the stress due to seismic bending forces (to be found at the base of the skirt by Eq. 9.3.17).

To make sure that the skirt thickness so found does not fail, following conditions are to be checked:

$$\sigma_z \text{ (tensile)} \leqslant f J \cos \alpha \qquad \ldots (10.2.3)$$

$$\sigma_z \text{ (compressive)} \leqslant 0.125 \, E \, (t/D) \cos \alpha \qquad \ldots (10.2.4)$$

Where, f = allowable stress of skirt material (usually at room temperature).

J = circumferential weld-joint efficiency factor.

= 1 (if made of single length, i.e., no circumferential joint)

= 0.7 (if double welded butt-joint with full penetration)

E = modulus of elasticity (Appendix A)

t = skirt thickness

D = skirt diameter

α = half the top angle of conical skirt

= 10° (maximum)

= 0° for cylindrical skirt.

The minimum thickness of skirt wall in corroded condition should not be less than 7 mm as per IS: 2825 − 1969. If the product of the skirt diameter (mm), thickness (mm), and temperature at the top of the skirt (in contact with vessel) above ambient (°C) exceeds 16×10^6 (mm² °C), account should be taken of the discontinuity stresses in both skirt and vessel induced by the temperature gradient in the upper section of the skirt. These stresses can be assessed as described in section 5.3.3.

In Eqs. 10.2.1 and 10.2.2 stress due to eccentric loads are neglected.

10 2.2 Design of skirt-bearing-plate and anchor-bolt

The bearing-plate at the base of the skirt is essential to increase the load-bearing contact area with the foundation, which has got low bearing capacity per unit area. Depending upon the mixes the allowable compressive strength of concrete foundation varies between 5.5 and 9.5 MN/m².

The bearing-plate, which is welded to the bottom of the skirt of the vessel, must be securely anchored to the concrete foundation by means of anchor bolts embedded in the concrete to prevent overturning from the bending moments induced by wind or seismic loads.

The maximum compressive stress between the bearing plate and the concrete foundation is given by:

$$\sigma_c \text{(max)} = \frac{W_{max}}{A} + \frac{(M_w \text{ or } M_s)}{Z} \qquad \ldots (10.2.5)$$

Where, W_{max} = maximum weight of the vessel (including liquid content and attachments),

A = area of contact between bearing plate and concrete foundation,

M_w = bending moment due to wind load,

M_s = bending moment due to seismic load,

Z = section modulus of area A.

The maximum compressive stress, σ_c (max.), must be less than the allowable bearing strength of the foundation.

The thickness of the bearing plate is determined by considering it as a uniformly loaded cantilever beam with σ_c (max.) the uniform load. The maximum bending moment for such a beam occurs at the junction of the skirt and the bearing plate, and is determined for unit circumferential width (i.e. $b = 1$) by the following equations:

$$M \text{(max)} = \sigma_c \, bl \, (l/2) = \frac{\sigma_c \, l^2}{2} \text{ (for } b = 1\text{)} \qquad \ldots (10.2.6)$$

Where, l = outer radius of bearing plate minus outer radius of the skirt.

The maximum stress in an elemental strip of unit width will be:

$$\sigma_{max} = \frac{6 \, M \text{(max)}}{b \, t_{bp}^2}$$

$$= \frac{3 \, \sigma_c \, l^2}{t_{bp}^2} \text{ (for } b = 1\text{)} \qquad \ldots (10.2.7)$$

Where, t_{bp} = thickness of the bearing-plate without gussets

Assuming $\sigma_{max} = f$ (allowable stress), and solving for t_{bp} from Eq. 10.2.7, gives

$$t_{bp} = l \sqrt{3 \, \sigma_c / f} \qquad \ldots (10.2.8)$$

If the calculated thickness of the bearing-plate after rounding off to the standard thickness is found 12 mm or less, a steel rolled-angle may be used as the bearing-plate of the skirt. (Fig. 10.2 a).

The thickness of the rolled-angle bearing-plate may be calculated from the following equation[1]:

$$\frac{3 \, \sigma_c \, (l - 1.7 \, t^2)}{t^2} < f \qquad \ldots (10.2.9)$$

---

[1] N. I. Taganov, "Fundamentals of Process Equipment Design", Academic Books, Delhi.

DESIGN OF SUPPORT FOR PROCESS VESSELS

Where, l = width of rolled angle,
t = thickness of rolled-angle.

If the calculated plate thickness is from 12-20 mm, a single-ring bearing plate may be employed. (Fig. 10.2 b).

(a) ROLLED-ANGLE BEARING PLATE

(b) SINGLE-RING BEARING PLATE WITH GUSSETS

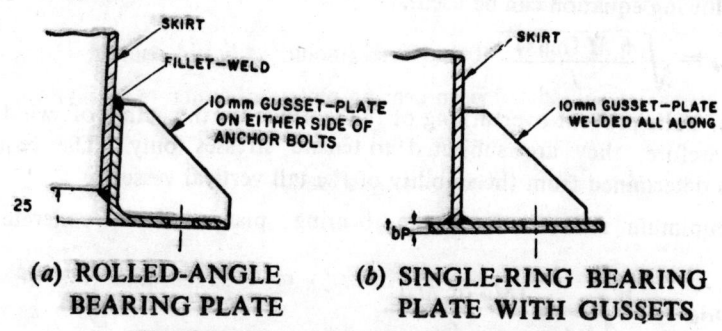

Fig. 10.2 Bearing plate for skirt support

If gussets are used to stiffen the bearing plates, the loading condition on the section of the plate between two gussets may be considered as a rectangular uniformly loaded plate with two opposite edges simply supported by the gussets, the third edge joined to the shell, and the fourth edge free. The deflections and bending moments in such case can be calculated applying the theory of strength of materials[2]. Table 10.1 gives the maximum bending moments for this case.

Table 10.1 Maximum Bending Moments in a Bearing Plate with Gussets

l/b	$M_x \begin{pmatrix} x = b/2 \\ y = l \end{pmatrix}$	$M_y \begin{pmatrix} x = b/2 \\ y = 0 \end{pmatrix}$
0	0	$-0.500\ \sigma_c\ l^2$
1/3	$0.0078\ \sigma_c\ b^2$	$-0.428\ \sigma_c\ l^2$
1/2	$0.0293\ \sigma_c\ b^2$	$-0.319\ \sigma_c\ l^2$
2/3	$0.0558\ \sigma_c\ b^2$	$-0.227\ \sigma_c\ l^2$
1	$0.0972\ \sigma_c\ b^2$	$-0.119\ \sigma_c\ l^2$
3/2	$0.123\ \sigma_c\ b^2$	$-0.124\ \sigma_c\ l^2$
2	$0.131\ \sigma_c\ b^2$	$-0.125\ \sigma_c$
3	$0.133\ \sigma_c\ b^2$	$-0.125\ \sigma_c$

b = gusset spacing (x direction),
l = bearing-plate outside radius minus skirt outside radius (y direction),
M_x = maximum bending moment at the outer edge mid point caused by deflection of the plane in the x-direction.
M_y = maximum bending moment at the junction of the skirt and bearing-plate caused by deflection of the plane in the y-direction.

[2] S. Timoshenko, "Theory of Plates and Shells", McGraw-Hill, New York.

It may be noted from Table 10.1 that for $l/b = 0$ (no gussets, i.e., gusset spacing, $b = \infty$) the bending moment reduces to Eq. 10.2.6, and the thickness of the bearing plate is determined by Eq. 10.2.8.

To determine the bearing-plate thickness from the bending moments given in Table 10.1, following equation can be used[3].

$$t_{bp} = \sqrt{\frac{6 M (max)}{f}} \qquad \ldots (10.2.10)$$

Anchor-bolts prevent overturning of the vessel from the action of wind or seismic forces, and therefore, they are subjected to tensile stresses only. The requirement of anchor-bolts is determined from the stability of the tall vertical vessel.

The minimum stress between the bearing plate and the concrete foundation will be,

$$\sigma_{min} = \frac{W_{min}}{A} - \frac{(M_w \text{ or } M_s)}{Z} \qquad \ldots (10.2.11)$$

Where, W_{min} = minimum weight of the empty vessel, i.e. without any internal attachments even.

If the value of σ_{min} is positive or zero, it is necessary to determine the so-called stability factor from the following equation[1]:

$$j = \frac{M_{weight}}{M_w \text{ (or } M_s)} = \frac{W_{min} R}{M_w \text{ (or } M_s)} \qquad \ldots (10.2.12)$$

Where, M_{weight} = moment of minimum weight of vessel,
R = moment arm for that weight of the vessel,
$= 0.42 D_o'$
and D_o' = outer diameter of bearing-plate.

If j is greater than 1.5, the vessel need not be anchored. In this case the bolts are used only for the fixation of the position of the vessel on the foundation.

On the other hand, if σ_{min} is negative, the vessel must be anchored to the concrete foundation by means of anchor bolts to prevent overturning owing to the bending moment induced by the wind or seismic load.

The approximate value of the load on one bolt may be determined from the following equation:

$$P_{bolt} = \sigma_{min} \frac{A}{n} \qquad \ldots (10.2.13)$$

where, P_{bolt} = load on one anchor bolt,
n = number of anchor bolts,
σ_{min} = stress determined by Eq. 10.2.11,
A = area of contact between bearing plate and foundation.

---

[3] L. E. Brownell and E. H. Young, "Process Equipment Design", John Wiley and Sons, Inc., New York.

DESIGN OF SUPPORT FOR PROCESS VESSELS

Once bolt load is determined from Eq. 10.2.13, the root area of the bolt can be found out by knowing the allowable stress value of the bolt material.

Design Example 10.1 : A cylindrical skirt support is to be designed for the vessel of Example 9.1 shown in Fig. 9.3. The material of construction for the skirt is decided to be IS : 2062—1962 Gr. St 42-W for which $f = 96$ MN/m^2 and $E = 2 \times 10^5$ MN/m^2. Determine the dimensions of the bearing-plate indicating if gussets are required. Evaluate also the requirement of anchor bolts.

Solution. It is to be noted that the tensile stress in the skirt will be maximum when the dead weight is minimum, i.e., the shell of the vessel is just erected and the shell is empty without any internal attachments. The compressive stress, on the other hand, is to be determined when the vessel is filled up with water for hydraulic test. Maximum wind load may be expected at any time and this factor is always to be considered.

The minimum weight of the vessel with two heads and shell will be :

$$W_{min} = \pi (D_i + t_a) t_a (H - 4) \gamma_s + 2 (7\,500)$$

where, $D_i = 1.972$ m

$t_a = 0.014$ m

$H = 20$ m

γ_s = specific weight (or weight density) of shell material
   = $7\,850 \times 9.81$ N/m^3
   = $77\,000$ N/m^3 or 77 kN/m^3

Substituting the values and expressing the weight in kN,

$W_{min} = (3.14)(1.986)(0.014)(16)(77) + 2(7.5)$
  $= 108 + 15$
  $= 123$ kN

$W_{max} = W_s + W_i + W_l + W_a$

where, W_s = weight of shell during test
   = 108 kN

W_i = weight of insulation
  = 42 kN

W_l = weight of water during test
  $= \dfrac{\pi D_i^2}{4} (H - 4)(9.81)$, kN
  $= (0.785)(1.972)(16)(9.81)$
  $= 244$ kN

W_a = weight of attachments
  = 71 kN

Therefore,

$W_{max} = 108 + 42 + 244 + 71$
   $= 465$ kN

Period of vibration at minimum dead weight is

$$T_{min} = 6.35 \times 10^{-5} (H/D)^{\frac{3}{2}} (W_{min}/ta)^{\frac{1}{2}}$$
$$= 0.185 \text{ s} < 0.5 \text{ s}$$

Hence,

$K_2 = $ a coefficient to determine wind load. (Eq. 9.3.9),
$= 1$

Period of vibration at maximum dead weight is given by,

$$T_{max} = 6.35 \times 10^{-5} (H/D)^{\frac{3}{2}} (W_{max}/t)^{\frac{1}{2}}$$
$$= 0.36 \text{ s} < 0.5 \text{ s}$$

Hence, $K_2 = 1$

From Eq. 9.3.9 the wind load is determined as follows:

$$P_w = K_1 K_2 p_w H D$$

For minimum weight condition, $D_o = 2$ m
For maximum weight condition, $D_o = 2.15$ m (insulated)
Hence, P_w (min) $= (0.7) (1) (1\ 000) (20) (2)$
$= 28\ 000$ N
and, P_w (max) $= (0.7) (1) (1\ 000) (20) (2.15)$
$= 30\ 100$ N

Minimum and maximum wind moments are computed by Eq. 9.3.11.

$$M_w \text{ (min)} = P_w \text{ (min)} \times \frac{H}{2}$$
$$= 28 \times 10 \text{ kJ}$$
$$= 280 \text{ kJ}$$

$$M_w \text{ (max)} = P_w \text{ (max)} \times \frac{H}{2}$$
$$= 30.1 \times 10 \text{ kJ}$$
$$= 301 \text{ kJ}$$

As the thickness of the skirt is expected to be small, assume

$D_i \approx D_o = 2$m

By Eq. 9.3.13

$$\sigma_{zwm} \text{ (min)} = \frac{4M_w \text{ (min)}}{\pi D^2 t}$$
$$= \frac{4 (0.280)}{3.14 (4) t} \text{ MN/m}^2$$
$$= 0.09/t \text{ MN/m}^2$$

$$\sigma_{zwm} \text{ (max)} = \frac{4 M_w \text{ (max)}}{\pi D^2 t}$$

DESIGN OF SUPPORT FOR PROCESS VESSELS

$$= \frac{4 (0.301)}{3.14 (4) t} \text{ MN/m}^2$$

$$= 0.096/t \quad \text{MN/m}^2$$

Minimum and maximum dead load stresses are calculated as follows :

$$\sigma_{zw} \text{ (min)} = \frac{W_{min}}{\pi d t}$$

$$= \frac{0.123}{3.14 (2) t} \text{ MN/m}^2$$

$$= 0.02/t \text{ MN/m}^2$$

$$\sigma_{zw} \text{ (max)} = \frac{W_{max}}{\pi D t}$$

$$= \frac{0.465}{3.14 (2) t} \text{ MN/m}^2$$

$$= 0.074/t \text{ MN/m}^2$$

By Eq. 10.2.1 maximum tensile stress without any eccentric load is computed as follows:

$$\sigma_z \text{ (tensile)} = \sigma_{zwm} \text{ (min)} - \sigma_{zw} \text{ (min)}$$

$$= 0.09/t - 0.02/t$$

$$= 0.07/t$$

Substituting, σ_z (tensile) $= f J$

$$= 96 \times 0.7$$

$$= 67.2 \text{ MN/m}^2$$

$J = 0.7$ for double-welded butt-joint for class 3 construction.

Equating,

$$67.2 = 0.07/t$$

$$t = \frac{0.07 \times 1\,000}{67.2} \text{ mm}$$

$$= 1.02 \text{ mm}$$

By Eq. 10.2.2 maximum compression load is computed as follows :

$$\sigma_z \text{ (compression)} = \sigma_{zwm} \text{ (max)} + \sigma_{zw} \text{ (max)}$$

$$= 0.096/t + 0.074/t$$

$$= 0.17/t$$

Substituting, σ_z (compressive) $= 0.125 \, E \, (t/D_o)$

$$= 0.125 \, (2 \times 10^5) \, (t/2)$$

$$= 1.25 \times 10^4 \, t$$

D—22

Equating,
$$1.25 \times 10^4 \, t = 0.17/t$$
$$\text{or, } t^2 = \frac{0.17}{1.25 \times 10^4}$$
$$\text{or, } t = 3.7 \times 10^{-3} \text{ m}$$
$$= 3.7 \text{ mm}$$

As per IS : 2825 — 1969 minimum corroded skirt thickness is 7 mm. Providing 1 mm corrosion allowance, a standard 8 mm thick plate can be used for skirt.

Design of skirt-bearing-plate :

By Eq. 10.2.5, maximum compressive stress between bearing-plate and foundation is :
$$\sigma_c = \frac{W_{max}}{A} + \frac{M_w}{Z}$$
Where, $W_{max} = 465$ kN
$A = \pi (D_o - l) l$
$D_o = $ outer diam. of skirt
$= 2$ m
$l = $ outer radius of bearing plate minus outer radius of skirt
$M_w = 301$ kJ
$Z = \pi R_m^2 \, l$
$R_m = (D_o - l)/2$

In calculating A, it is assumed that the bearing-plate's inside diameter is not much less than the inside diameter of skirt. Inside extension of bearing plate is made to reduce compressive load concentration on foundation, it does not take bending moment.

Substituting the values in the above equation,
$$\sigma_c = \frac{0.465}{3.14 (2 - l) \, l} + \frac{0.301 \times 4}{3.14 (2 - l)^2 \, l}$$

From section 10.2.2, the allowable compressive strength of concrete foundation varies from 5.5 to 9.5 MN/m². Assuming, $\sigma_c = 5.5$ and substituting in the above relationship, l is found to be equal to 35 mm. As the required width of the bearing-plate is very small, a 100 mm width is selected (i.e. $l = 0.1$ m).

By Eq. 10.2.7, the thickness of the bearing plate is calculated as follows :
$$t_{bp} = l\sqrt{3 \, \sigma_c / f}$$
Where, $l = 100$ mm
$\sigma_c = $ maximum compressive load calculated by Eq. 10.2.5 for $l = 0.1$ m
$= 1.85$ MN/m²
$f = $ allowable stress
$= 96$ MN/m²

Substituting,
$$t_{bp} = 100\sqrt{(3 \times 1.85/96)}$$
$$= 25 \text{ mm}$$

Bearing plate thickness of 25 mm is required. As the plate thickness required is larger than 20 mm, gussets may be used to reinforce the plate.

DESIGN OF SUPPORT FOR PROCESS VESSELS

From Table 10.1 and for $l/b = 1$,
$$M(\max) = M_y = -0.119\, \sigma_c\, l^2$$
$$= -0.119 \times 1.85\,(0.1)^2,\ \text{MJ}$$
$$= -2.2 \times 10^{-3}\ \text{MJ}$$

By Eq. 10.2.10
$$t_{bp} = \sqrt{\frac{6\,M(\max)}{f}}$$
$$= \sqrt{\frac{6\,(2.2 \times 10^{-3})}{96}},$$
$$= 1.18 \times 10^{-2},\ \text{m}$$

If gussets are used at 100 mm spacing, bearing plate thickness of 12 mm will be sufficient. Therefore, rolled-angle bearing plate (Fig. 10.2 a) of size 100 mm × 100 mm × 12 mm with 62 gussets may be used. If number of the gussets are reduced, thicker plate will be required.

By Eq. 10.2.11,
$$\sigma_{min} = \frac{W_{min}}{A} - \frac{M_w}{Z}$$
$$= \frac{0.123}{3.14\,(2-0.1)\,(0.1)} - \frac{0.280}{3.14\,(2-0.1)^2\,(0.1)},\ \text{MN/m}^2$$
$$= 0.206 - 0.203,\ \text{MN/m}^2$$
$$= 0.003,\ \text{MN/m}^2$$

By Eq. 10.2.12,
$$j = \frac{W_{min}\,R}{M_w} = \frac{0.123\,(0.42)\,(2 + 0.2)}{0.280}$$
$$= 0.4$$

As this value is less than 1.5, the vessel will not be steady by its own weight. Therefore, anchor bolts are to be used.

By Eq. 10.2.13,
$$P_{bolt}(n) = \sigma_{min}\,A$$
$$= 0.003 \times 3.14\,(2 - 0.1)\,(0.1),\ \text{MN}$$
$$= 1.82 \times 10^{-3},\ \text{MN}$$

If hot rolled carbon steel is selected for bolts, Table 7.5 gives $f = 57.3\ \text{MN/m}^2$.
Therefore, $(a_r n)\,f = n\,P_{bolt} = 1.82 \times 10^{-3}$, MN

or $a_r\,n = \dfrac{1.82 \times 10^{-3}}{57.3}$, m^2

$$= 3.15 \times 10^{-5}\ \text{m}^2$$
$$= 31.5\ \text{mm}^2$$

Where,
a_r = root area of a bolt
n = no of bolts.

For M 12 × 1.5 bolts, $a_r = 63\ \text{mm}^2$. This indicates that one such anchor-bolt is more than sufficient for the purpose. But, as the wind may blow from any side, 8 such bolts are to be used equally distributed.

10.3 DESIGN OF LUG OR BRACKET SUPPORT

Brackets or lugs are widely used for all types of vessels. These are used to support vertical vessels of smaller height (subjected to minor wind loads). Lugs are welded to the vessel and rest on columns or beams as shown in Fig. 10.3.

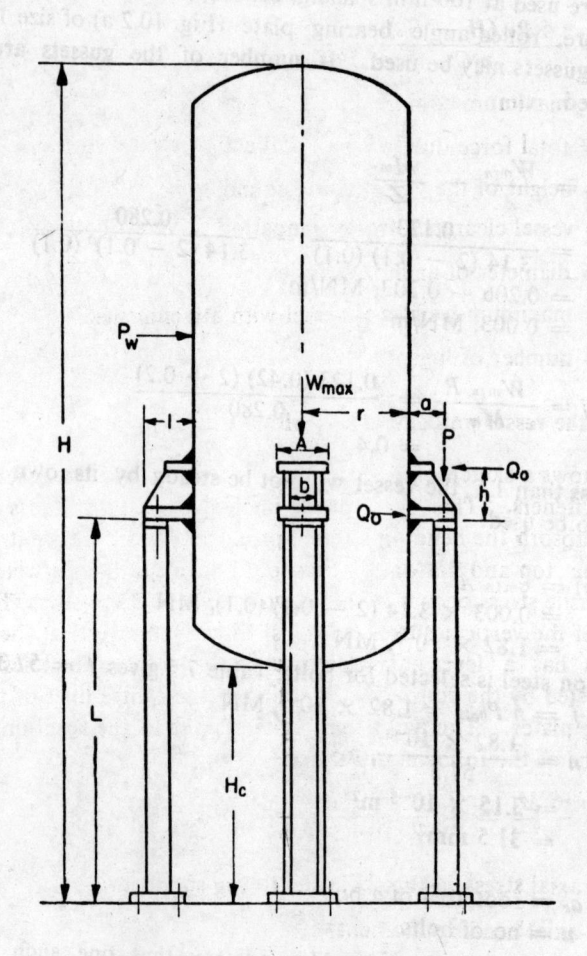

Fig. 10.3 Vessel on lug supports

The advantages of lug or bracket supports are, — they are easily constructed and attached to the vessels with a minimum length of the weld seam. They can absorb diametrical expansions (provided they are equipped with a device for sliding) and are easily levelled. The disadvantage with this type of support is that as a result of the eccentricity, the compressive, tensile and shear stresses are induced in the vessel wall. Therefore, in thin-walled vessel this type of support requires reinforcement of the wall with backing plate.

The maximum load on the lug is the compression load on downwind side due to wind and weight of the vessel and its contents. Therefore the stresses on the downwind side are the determining factor for design of the supports. The maximum total compression load on one lug will be[3] (Fig. 10.3):

$$P = \frac{4 P_w (H - H_c)}{n C} + \frac{W_{max}}{n} \quad \ldots (10.3.1)$$

Where, P = maximum compression load per lug,

P_w = total force due to wind load acting on the vessel,

H = height of the vessel above foundation,

H_c = vessel clearance from foundation to vessel bottom,

C = diameter of anchor bolt circle,

W_{max} = maximum weight of vessel with attachments.

n = number of lugs.

10.3.1 Stresses in the vessel wall due to lugs with horizontal plates

Fig. 10.3 shows a sketch of a vessel supported on 4 lugs, each lug having two horizontal-plant stiffeners. This type of lug imparts the axial stiffness and strength of the cylindrical shell to absorb the bending stresses produced by the concentrated loads of the supports. Both the top and bottom plates should have continuous welds as the maximum compressive and tensile stress occurs in these two plates, respectively. These welds and the intermittent welds of the vertical gussets to the shell carry the vertical shear load. The load, P, on the column has a lever arm, a, measured to the mean thickness of the shell plate. This moment is resisted by the couple, $Q_o h$, acting at the centre lines of the top and bottom horizontal stiffening plates. The axial bending stress due to the reaction of the lug may be determined by means of the following equation[3]:

$$\sigma_{zl} = \frac{6 M}{t^2} = \pm \frac{\beta^3 P a r^2}{2 (1 - \mu^2) A h} \quad \ldots (10.3.2)$$

where, σ_{zl} = axial stress in the shell due to lug support,

M = axial bending moment,

t = vessel wall thickness,

P = maximum compression load on a lug,

a = lever arm for load, P, from bolt circle,
r = mean radius of the shell,
A = width of compression plate,
h = gusset or rib height,
μ = Poisson's ratio (0.3 for steel),

$$\beta = \sqrt[4]{\frac{3(1-\mu^2)}{r^2 t^2}}$$

If the horizontal plates are not used or if the bottom plate is not welded to the shell, Eq. 10.3.2 cannot be applied for determining axial bending stresses. lugs are welded through backing plate, it should be strong enough to carry the load and in that case Ah of Eq. 10.3.2 will represent the backing plate area. Amount of welding between shell and backing plate should be enough to carry the vertical shear load.

If the vessel is subjected to internal or external pressure, axial stress due to pressure will be added to the axial bending stress and these two stresses combined should not exceed the allowable tensile strength of the shell material. Therefore,

$$f \geqslant \sigma_{zp} + \sigma_{zl} \qquad \ldots (10.3.3)$$

where, $\sigma_{zp} = \dfrac{pD}{4t}$

For $p = 0$,

$$t \geqslant 1.76 \left(\frac{Pa}{2Ahf}\right)^{\frac{2}{3}} r^{\frac{1}{3}} \qquad \ldots (10.3.4)$$

10.3.2 Thickness of horizontal and gusset plates

The procedure of calculation of the thickness of bottom horizontal plate of lug is essentially the same as that of calculating the compression-plate thickness of external chairs of skirt support[3]. For the case in which 'a' is assumed to be equal to 'l' (Fig. 10.3), the maximum bending moments M_y and M_x are given by :

$$M_y = \frac{P}{4\pi}(1 - \gamma_1) \qquad \ldots (10.3.5)$$

$$M_x = \frac{P}{4\pi}(\mu + \gamma_2) \qquad \ldots (10.3.6)$$

Where, M_y = maximum bending moment along radial axis,
M_x = maximum bending moment along circumferential axis,
P = maximum compression load on a lug,
μ = Poisson's ratio (0.3 for steel),
γ_1, γ_2 = constants from Table 10.2.

Table 10.2 Constants for Moment Calculation[3]

b/l	1.0	1.2	1.4	1.6
γ_1	0.555	0.350	0.211	0.125
γ_2	0.135	0.115	0.085	0.057

Between M_y and M_x whichever is larger will be the controlling bending moment to determine the thickness by the following equation :

$$t_{hp} = \sqrt{\frac{6 M \text{ (max)}}{f}} \qquad \ldots(10.3.7)$$

where, t_{hp} = thickness of horizontal plate,
M (max) = maximum bending moment,
f = allowable stress of the plate material.

If the gusset height is small and the ratio h/r_g is less than 60, the gusset thickness is determined from the relation of straight compression, i.e.

$$t_g = \frac{P/2}{f_c \, l} \qquad \ldots(10.3.8)$$

Where, r_g = least radius of gyration for gusset,
$r_g^2 = t_g^2/12$
t_g = gusset plate thickness,
h = height of gusset plate,
l = width of gusset plate,
P = concentric load for two gussets,
f_c = allowable compressive stress
≈ 100 MN/m² (maximum).

If h/r_g is larger than 60, column equation is to be used to determine f_c.

10.3.3 Design of column supports for lugs

The column support for a lug can be considered as a column under concentric axial load, and the allowable compressive stress is related to the radius of gyration of the column by the following equation :

$$f_c = \frac{120}{1 + \dfrac{L^2}{18\,000 \, r_g^2}} \qquad \ldots(10.3.9)$$

where, L = length of column, m
r_g = least radius of gyration of column, m
f_c = allowable column stress in column, MN/m²

Eq. 10.3.9 is applicable[3] for $\frac{l}{r_g}$ between 60 and 200. The required cross-sectional area, X, of the column for axial compression load, P, is given by :

$$X \geqslant \frac{P}{f_c} \qquad ...(10.3.10)$$

When, f_c is obtained from Eq. 10.3.9.

As the column is attached at a distance 'a' from the vessel wall, this produces eccentric load causing additional stress in the column support. This stress is given as :

$$f_{ec} = \frac{Pa}{Z} \qquad ...(10.3.11)$$

where, Z = section modulus of column.

The bending stress in the column due to wind load is given by :

$$f_{bend} = \frac{(P_w/n)(L/2)}{Z} \qquad ...(10.3.12)$$

When columns are subjected to direct loads and bending produced by eccentric loads simultaneously, following condition is required to be satisfied.

$$\frac{P/X}{f_c} + \frac{(Pa/Z) + (P_w L/2 \, nZ)}{f_t} \leqslant 1 \qquad ...(10.3.13)$$

where, f_c = allowable column stress,

f = allowable flexural stress.

The bearing plate for the column support is designed in the same way as that for the skirt support.

10.4 DESIGN OF SADDLE SUPPORTS

General practice of supporting horizontal cylindrical vessels is by means of saddle supports as shown in Fig. 10.4, although no rigorous analytical treatment of them is available. The horizontal vessels when resting on the saddles behave as a beam. Number of saddles may be two or more. Following is the analysis of stress produced in the wall when a vessel rests on two saddle supports[4]. The included angle, θ, of a saddle support should not normally be less than 120°. This limitation, which is imposed by most codes of practice, is an empirical one based on experience of large vessels[5].

---

[4] L. P. Zick, "Stresses in large horizontal cylindrical pressure vessels on two saddle supports", Welding Journal Research Supplement, September, 1951.

[5] IS : 2825—1969.

DESIGN OF SUPPORT FOR PROCESS VESSELS

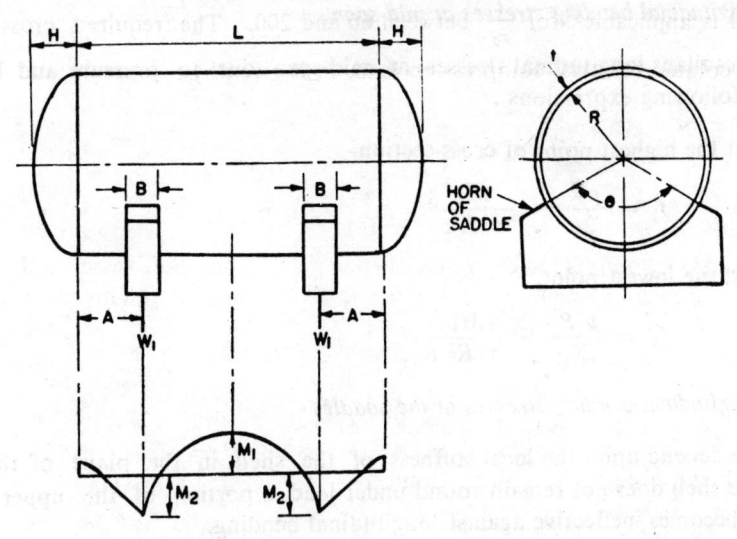

Fig. 10.4 Cylindrical shell acting as beam over saddle supports

10.4.1 Longitudinal bending moments in the vessel shell

The bending moment diagram for the shell is shown in Fig. 10.4. From this figure it is clear that there are two maximum bending moments over the supports and at the centre of the span. The bending moments are given by the following equations:

(a) At mid-span

$$M_1 = \frac{W_1 L}{4} \left[\frac{1 + \frac{2(R^2 - H^2)}{L^2}}{1 + \frac{4H}{3L}} - \frac{4A}{L} \right] \qquad \ldots (10.4.1)$$

(b) At supports

$$M_2 = -W_1 A \left[1 - \frac{1 - \frac{A}{L} + \frac{(R^2 - H^2)}{2AL}}{1 + \frac{4H}{3L}} \right] \qquad \ldots (10.4.2)$$

The positive bending moments from these expressions will cause tension at the lowest point of the shell cross-section. Nomenclature are to be seen from Fig. 10.4.

D—23

10.4.1.1 Longitudinal bending stresses at mid-span

The resultant longitudinal stresses at mid-span due to pressure and bending are given by the following expressions:

(a) At the highest point of cross-section,

$$f_1 = \frac{pR}{2t} - \frac{M_1}{\pi R^2 t} \qquad \ldots (10.4.3)$$

(b) At the lowest point,

$$f_1' = \frac{pR}{2t} + \frac{M_1}{\pi R^2 t} \qquad \ldots (10.4.4)$$

10.4.1.2 Longitudinal bending stresses at the saddles

These depend upon the local stiffness of the shell in the plane of the supports, because, if the shell does not remain round under load, a portion of the upper part of its cross section becomes ineffective against longitudinal bending.

When the supports are near the end of the vessel, so that $A < R/2$, the stiffness of the ends is enough to maintain a circular cross-section. Such shells are said to be stiffened by the ends.

The resultant longitudinal stresses due to pressure and bending are given by:

(a) At the highest point of the cross-section in shells which remain round in the plane of the support,

$$f_2 = \frac{pR}{2t} - \frac{M_2}{K_1 (\pi R^2 t)} \qquad \ldots (10.4.5)$$

(b) At the lowest point of the cross-section,

$$f_2' = \frac{pR}{2t} + \frac{M_2}{K_2 (\pi R^2 t)} \qquad \ldots (10.4.6)$$

Values of K_1 and K_2 are given in Table 10.3.

Table 10.3 Values of Factors K_1 and K_2

Condition	Saddle Angle (θ)	K_1	K_2
Shell stiffened by end or rings (i.e. $A < R/2$ or rings provided)	120	1	1
	150	1	1
Shell unstiffened by end or rings (i.e. $A > R/2$ and no rings provided)	120	0.107	0.192
	150	0.161	0.279

It is to be noted that the induced tensile stresses should not exceed the allowable design stress, (fJ). Similarly, compressive stresses should not numerically exceed the allowable design stress or $E\,t/16\,R$ whichever is less.

10.4.2 Tangential shearing stresses

The load is transferred from the unsupported part of the shell to the part over the supports by tangential shearing stresses which vary with the local stiffness of the shell.

Case 1: Shell not stiffened by vessel end ($A > R/2$)

The maximum tangential shearing stress is given by:

$$q = \frac{K_3\,W_1}{R\,t}\left(\frac{L - 2A - H}{L + H}\right) \qquad \ldots (10.4.7)$$

Eq. 10.4.7 is not applicable if $A > L/4$, but such proportions are unusual.

The value of K_3 depends on the presence or absence of supporting rings and on the saddle angle, θ. It is given in Table 10.4.

The thickness of the local backing plates should not be included in Eq. 10.4.7.

Case 2: Shell stiffened by end of vessel ($A < R/2$)

In this case there are shearing stresses in both the shell and the vessel end, given by:

$$q = \frac{K_3\,W_1}{R\,t} \text{ in the shell} \qquad \ldots (10.4.8)$$

$$q_e = \frac{K_4\,W_1}{R\,t_e} \text{ in the end} \qquad \ldots (10.4.9)$$

The values of K_3 and K_4 depend on the width, B, of the saddle and on the saddle angle, θ. They are given in Table 10.4.

It has been suggested[4] that the maximum allowable values of the tangential shear stresses should be:

$$q = 0.8\,f \text{ in the shell}$$

and $\quad q_e = (1.15\,f - f_n)$ in the end

Where, $\quad f_n = p\,D_o\,C/2\,t$

in which C is the shape factor for domed ends given in section 4.2.3.

These high values can be allowed because the resultant combined stresses are local in character and will be relieved in practice by prestressing due to local yielding.

Table 10.4 Values of Factors[5] K_3 and K_4

Condition	Saddle Angle (θ)	K_3	K_4
$A > R/2$ and shell unstiffened by rings	120	1.171	...
	150	0.799	...
$A > R/2$ and shell stiffened by rings in plane of saddles	120	0.319	...
	150	0.319	...
$A > R/2$ and shell stiffened by rings adjacent to saddles	120	1.171	...
	150	0.799	...
Shell stiffened by end of vessel $\left. \begin{array}{l} B < A \leqslant R/2 \\ \frac{H}{2} < A < H \end{array} \right\}$	120	0.880	0.401
	150	0.485	0.297
	120	0.880	0.880
	150	0.485	0.485

10 4.3 Circumferential stresses

The circumferential stresses are calculated from the following semi-empirical correlations[5], for a shell not stiffened by rings:

(a) At the lowest point of the cross-section,

$$f_3 = -\frac{K_5 W_1}{t(B + 10 t)} \quad \ldots (10.4.10)$$

(b) At the horn of the saddle,

if $L/R > 8$, $f_4 = -\frac{W_1}{4 t (B + 10 t)} - \frac{3 K_6 W_1}{2 t^2} \quad \ldots (10.4.11)$

if $L/R < 8$, $f_4 = -\frac{W_1}{4 t (B + 10 t)} - \frac{12 K_6 W_1 R}{L t^2} \quad \ldots (10.4.12)$

These stresses may be reduced if necessary by welding a reinforcing backing plate to the shell between it and saddle as shown in Fig. 10.5. If the width of this plate is not less than $(B + 10 t)$ and it subtends an angle not less than $(\theta + 12)$ degrees at the centre of the cylinder, the reduced stresses at the edge of the saddle can be obtained by substituting $(t + t_1)$, the combined thickness of shell and reinforcing backing plate, for t in the above equations.

Positive values denote tensile stresses and negative values denote compression.

The numerical values of the circumferential stresses found as above should not exceed 1.25 times the allowable working stress in tension.

Values of K_5 and K_6 are given below:

$K_5 = 0.760$ for $\theta = 120°$
$ = 0.673$ for $\theta = 150°$

DESIGN OF SUPPORT FOR PROCESS VESSELS 181

Table 10.5 Values of K_6

A/R	$\theta = 120°$	$\theta = 150°$
0 — 0.5	0.013	0.007
0.6	0.018	0.010
0.7	0.030	0.017
0.8	0.034	0.021
0.9	0.047	0.028
1.0	0.052	0.031
1.1 — 3.0	0.055	0.033

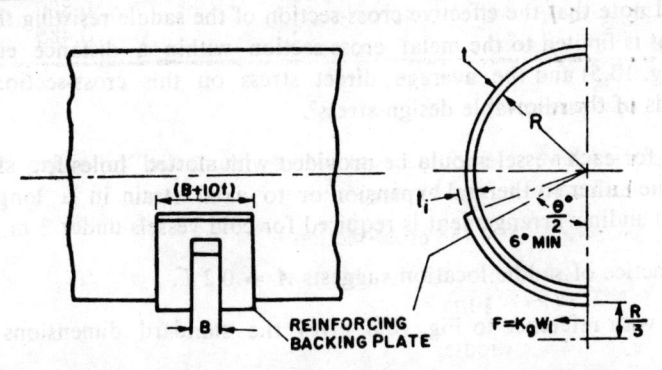

Fig. 10.5 Saddle support and reinforcing backing plate

10.4.4 Ring stiffeners

Stiffeners are attached to the shell (usually inside the shell) over the saddle to alleviate the load on the shell.

Ring stiffener is designed from the following correlation:

$$f = -\frac{K_7 W_1}{A_r} \pm \frac{K_8 W_1 R}{Z} \qquad \ldots (10.4.13)$$

Where,
f = allowable compressive stress,
A_r = cross-sectional area of the stiffening ring, (thickness × width for rectangular cross-section),
Z = section modulus of ring cross-section.

The values of K_7 and K_8 as a function of saddle angle, θ, are given in Table 10.6.

Table 10.6 Values[1] of K_7 and K_8

Saddle Angle (θ)	K_7	K_8
120	0.056 0	0.052 8
150	0.021 0	0.031 6

10.4.5 Design of saddles

The saddle should be strong enough to withstand the forces imposed by the vessel.

The width B of steel saddles should not be less than $10\,t$. It should be increased if the circumferential stresses exceeds acceptance limit.

The horizontal component of all radial loads may be determined by the following equation.

$$F = K_9 W_1$$

Where,
$K_9 = 0.204$ for $\theta = 120°$
$ = 0.260$ for $\theta = 150°$

One should note that the effective cross-section of the saddle resisting this horizontal force component is limited to the metal cross-section within a distance equal to $R/3$ below the shell (Fig. 10.5) and the average direct stress on this cross-section should be limited to two-thirds of the allowable design stress[5].

One saddle for each vessel should be provided with slotted holes for sliding, when large movements due either to thermal expansion or to axial strain in a long vessel are expected. No such sliding arrangement is required for cold vessels under 3 m long.

General practice of saddle location suggests $A = 0.2\,L$.

Table 10.7, with reference to Fig. 10.6, gives the standard dimensions for saddles used in practice.

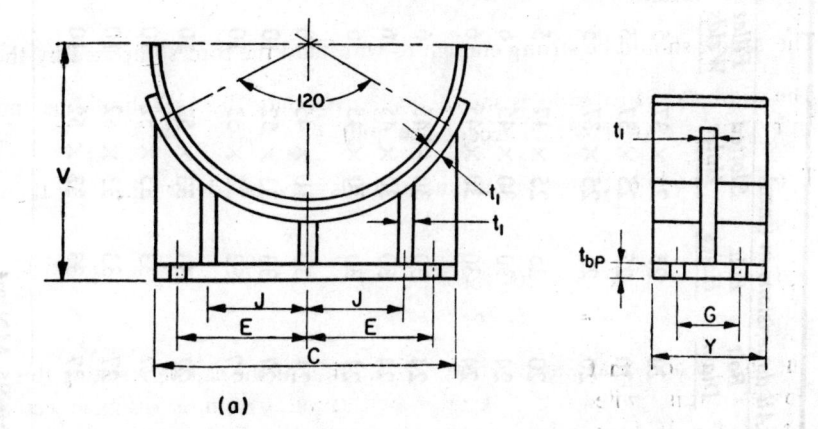

(a) FOR VESSEL DIA. 1.2 m AND UNDER

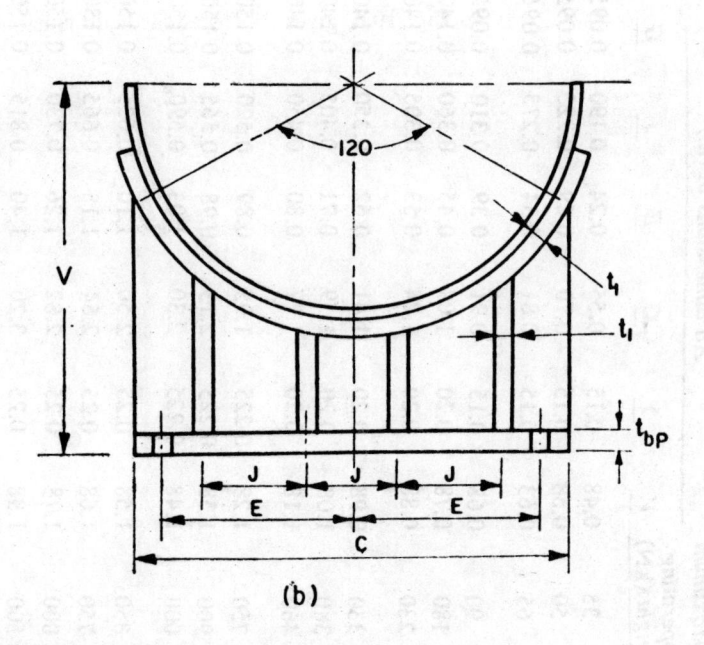

(b) FOR VESSEL DIA. OVER 1.2 m

Fig. 10.6 Standard construction of steel saddles (Table 10.7)

Table 10.7 Standard Dimensions for Saddles (Fig. 10.6)

Vessel	Maximum operating weight (kN)	All dimensions in (m)							All dimensions in (mm)				
Diam. (m)		V	Y	C	E	J	G	t_{br}	t_1	Bolt Diam.	Bolt Holes	Slotted Holes	Fillet Welds
0.6	35	0.48	0.15	0.55	0.24	0.190	0.095	6	5	20	25	25 × 37	5
0.8	50	0.58	0.15	0.70	0.29	0.225	0.095	8	5	20	25	25 × 37	5
0.9	65	0.63	0.15	0.81	0.34	0.275	0.095	10	6	20	25	25 × 37	5
1.0	90	0.68	0.15	0.91	0.39	0.310	0.095	11	8	20	25	25 × 37	5
1.2	180	0.78	0.20	1.09	0.45	0.360	0.140	12	10	24	30	30 × 45	6
1.4	230	0.88	0.20	1.24	0.53	0.305	0.140	12	10	24	30	30 × 45	6
1.6	330	0.98	0.20	1.41	0.62	0.350	0.140	12	10	24	30	30 × 45	6
1.8	380	1.08	0.20	1.59	0.71	0.405	0.140	12	10	24	30	30 × 45	6
2.0	460	1.18	0.20	1.77	0.80	0.450	0.140	12	10	24	30	30 × 45	6
2.2	750	1.28	0.225	1.95	0.89	0.520	0.150	16	12	24	30	30 × 45	10
2.4	900	1.38	0.225	2.13	0.98	0.565	0.150	16	12	27	33	33 × 52	10
2.6	1 000	1.48	0.225	2.30	1.03	0.590	0.150	16	12	27	33	33 × 52	10
2.8	1 350	1.58	0.25	2.50	1.10	0.625	0.150	16	12	27	33	33 × 52	10
3.0	1 750	1.68	0.25	2.64	1.18	0.665	0.150	16	12	27	33	33 × 52	10
3.2	2 000	1.78	0.25	2.82	1.26	0.730	0.150	16	12	27	33	33 × 52	10
3.6	2 500	1.98	0.25	3.20	1.40	0.815	0.150	16	12	27	33	33 × 52	10

Note: Continuous fillet welds all round. Maximum allowable working stress 95 MN/m².

CHAPTER 11

DESIGN OF THICK WALLED HIGH PRESSURE VESSELS

11.1 INTRODUCTION

The applicability of standard equations given in Chapter 3 for determining shell thickness of unfired pressure vessels is limited to pressure rating upto 20 MN/m² and D_o/D ratio not exceeding 1.5. To-day many chemical process are required to operate at much higher pressure ratings. For example, Haber Bosch process of making Ammonia had become a commercial proposition calling for pressure vessels which can withstand 10 to 100 MN/m² pressure. Hydrogenation of coal, Methanol synthesis and Urea reactors also called for high pressure vessels of 10 to 30 MN/m². So also atomic power plants designed for about 20 to 30 MN/m². The latest technology of manufacture of polyethylene in the modern "Plastic Age" need pressure vessels withstanding 150 MN/m² pressure. Therefore, pressure vessels comprise of either a thin or thick cylinder with closures made of hemispheres or torispherical or semiellipsoidal dished ends, flat covers, etc. The vessel will be called thick walled if D_o/D_i exceed 1.5 or pressure exceeds 20 MN/m². In such cases following analysis is to be followed.

11.2. STRESSES IN A THICK CYLINDER

In case of thin shell under internal pressure, the variation of stress from inner to outer surface is not considerable. When the wall thickness is relatively large, the variation of stress from inside to outside surface of the cylinder will be very large and hence the equations given in Chapter 3 will not satisfy the design conditions. The variation of stress as given by Lame's solution for thick cylinder is given by the following equations :

$$\sigma_z = \frac{p_i D_i^2 - p_o D_o^2}{D_o^2 - D_i^2} \qquad \ldots (11.2.1)$$

$$\sigma_r = \frac{p_i D_i^2 - p_o D_o^2}{D_o^2 - D_i^2} - \frac{(p_i - p_o) D_i^2 D_o^2}{D^2 (D_o^2 - D_i^2)} \qquad \ldots (11.2.2)$$

$$\sigma_\theta = \frac{p_i D_i^2 - p_o D_o^2}{D_o^2 - D_i^2} + \frac{(p_i - p_o) D_i^2 D_o^2}{D^2 (D_o^2 - D_i^2)} \qquad \ldots (11.2.3)$$

where, D = any diameter where stress is to be evaluated.

The principal stresses can be expressed by the following equations :

Most of the practical vessel applications are falling under internal pressure only, where $p_o = 0$. Under this condition, the stresses experienced by the inner surface of a cylinder are :

$$\sigma_{r_i} = -p_i \qquad \ldots (11.2.4)$$

$$\sigma_{z_i} = p_i/(K^2 - 1) \qquad \ldots (11.2.5)$$

$$\sigma_{\theta_i} = p_i (K^2 + 1)/(K^2 - 1) \qquad \ldots (11.2.6)$$

where, $\quad K = D_o/D_i$

The stresses on the outer surface are :

$$\sigma_{r_o} = 0 \qquad \ldots (11.2.7)$$

$$\sigma_{z_o} = \sigma_{z_i} \qquad \ldots (11.2.8)$$

$$\sigma_{\theta_o} = 2 p_i/(K^2 - 1) \qquad \ldots (11.2.9)$$

From these equations it is clear that, the stress distribution in the cylinder follows hyperbolically for σ_θ. Its magnitude is maximum at the inner surface and minimum at the outer surface. σ_z is constant throughout the cross-section.

The maximum shear stress at any point in the cylinder is equal to one half the algebraic difference of Max. and Min. principal stresses at that point[1].

$$\tau = \frac{\sigma_\theta - \sigma_r}{2} = \frac{(p_i - p_o) D_i^2}{(D_o^2 - D_i^2)} \frac{D_o^2}{D^2} \qquad \ldots (11.2.10)$$

The above equation shows that the shear stress is maximum on the inner surface.

It is to be noted that it is not only σ_θ, which is responsible for the failure of a thick walled cylinder. The three principal stresses perpendicular to each other produce most intricate stress condition in a cylindrical wall, which can only be approximated by means of mathematics.

11.3 THEORIES OF ELASTIC FAILURE[2]

For a given material elastic failure is considered to occur when the elastic limit of the material is reached. Beyond this limit the specimen is permanently deformed or ruptured. Of the various theories developed to account for elastic failure, the following four theories are of special interest and failures will occur when,

1. The maximum principal stress equals the stress at the elastic limit under simple tension.
2. The maximum strain equals the strain at the elastic limit under simple tension.
3. The maximum shear stress reaches a critical value.
4. The strain energy per unit volume reaches a critical value.

---

[1] John F. Harvey, "Pressure Vessel Design", East-West Press Pvt. Ltd., New Delhi.

[2] L. E. Brownell and E. H. Young, "Process Equipment Design", John Wiley and Sons. Inc., New York.

11.3.1. Maximum principal stress theory

The failure is considered to occur when any of σ_r, σ_θ, or σ_z reaches elastic limit which is taken as yield point of material, σ_y.

$$\sigma_{\theta \ (max)} = \sigma_y = p_i (K^2 + 1)/(K^2 - 1) \qquad \ldots (11.3.1)$$

The above equation from Lame's solution is not theoretically correct as it does not include Poisson's ratio, μ, in it, and in many cases it has been reported that the equation resulting from maximum strain containing μ gives better understanding with the experimental value.

11.3.2 Maximum strain theory

In this case the failure will occur when the strain set up in the material reaches the strain at elastic limit by the induced stress.

$$\sigma_{\theta \ (max)} = \sigma_y = p_i \left[\frac{(1-\mu) + (1+\mu) K^2}{(K^2 - 1)} \right] \qquad \ldots (11.3.2)$$

11.3.3 Maximum strain energy theory

The strain energy is the mechanical energy absorbed by a body being stressed within the elastic limit. According to this theory, the failure will occur when the strain energy accumulated in the material is stressed up to elastic limits.

$$\sigma_y = \frac{p_i \sqrt{6 + 10 K^4}}{2 (K^2 - 1)} \qquad \ldots (11.3.3)$$

11.3.4 Maximum shear theory

The failure occurs when the maximum shear stress equals the shear stress set up in the material at elastic limit i.e. yield point of the materials.

$$\tau = \tfrac{1}{2} (\sigma_\theta - \sigma_r) = \tfrac{1}{2} \sigma_y$$

$$\text{or } \sigma_y = \frac{2 K^2}{(K^2 - 1)} p_i = 2 \tau_{(max)} \qquad \ldots (11.3.4)$$

As per Manning's discussion on the strength of high pressure cylinder, the most satisfactory result on failure is given by the equation,

$$\frac{\sigma_y}{\sqrt{3}} = \frac{K^2}{(K^2 - 1)} p_i$$

$$\text{or } \sigma_y = \frac{\sqrt{3} K^2}{K^2 - 1} p_i \qquad \ldots (11.3.5)$$

Eq. 11.3.5 is the modification of Eq. 11.3.4, in which $\frac{\sigma_y}{2}$ is substituted by $\sigma_y/\sqrt{3}$.

The equation $\frac{\sigma_y}{p_i} = \frac{\sqrt{3} K^2}{(K^2 - 1)}$ is the elastic break down by maximum shear theory.

$$p_i = \frac{1}{\sqrt{3}} \sigma_y \left(\frac{K^2 - 1}{K^2} \right)$$

$$= 0.58 \frac{(K^2 - 1)}{K^2} \sigma_y \qquad ...(11.3.6)$$

It is to be noted carefully that for design purpose the value of σ_y in Eqs. 11.3.1 through 11.3.6 must be divided by a suitable factor of safety.

11.4 LIMITATIONS OF MONOBLOCK HIGH PRESSURE VESSEL CONSTRUCTION WITH ORDINARY STEELS

Vessels made either from (1) thick plate or by (2) forging out of ingot are known as monoblock construction. It is easily possible to bend 50 mm thick plates by cold bending and then to weld the seam. Hot bending could be possible for plate thicknesses upto 150 to 175 mm; but it becomes increasingly difficult for bending thicknesses above that. It is also very difficult to procure thicker plates with sound and uniform metallurgical properties from the steel mills, and welding of these plates becomes very difficult.

With regard to forged construction, since it is made out of ingot, the steel mills need to produce ingots about two times more than the required weight of the shell ring and also need huge forging presses. Moreover, it is difficult to obtain material with uniform metallurgical properties throughout the section.

With above discussion if Eq. 11.3.6 is considered, it shows that with K keeps increasing, the term $(K^2 - 1)/K^2$ approaches to unity and that internal pressure and yield stress of the steel become practically in direct proportion. This means that beyond a given ratio of diameters (about $K = 3$) the internal pressure cannot be increased even by further increase of wall thickness of the cylinder, and the realization of this fact gives rise to three methods for the development of high pressure vessel constructions.

(i) By increasing the yield strength of material i.e., by using Ti steel whose properties can be improved by quenching and tempering. This gives the venue for material development.

(ii) By changing the stress distribution in the thick cylinder wall. This gives rise to multilayer vessel wall construction and the method of "Autofrettaging" the monoblock shell.

or, (iii) By applying both the methods together.

11.4.1 Monoblock vessel with high strength steels

For the reasons discussed in the earlier section, the attempt has been made to use high strength steel like Ti steel. Out of the 3 methods for improving high pressure vessel construction, it is very difficult to say which will be more economical. Everything depends on the facilities available, availability of proper material at reasonable price and the use of the vessel in the particular services.

DESIGN OF THICK WALLED HIGH PRESSURE VESSELS

The yield strength of the material can be increased by alloying the steel with Cr, Mo, V, Ni, etc. and also by heat treatment i.e. by quenching and tempering. The welding and the feasibility of heat treatment depend on the dimensions of the equipment and the furnace capacity available. If the thickness becomes high, the metallurgical quality after heat treatment will not be sound. Yet this type of forged heat treated vessels are in manufacture in Czechoslovakia, West Germany, Japan, and some other industrially advanced countries. In India the manufacture of such vessels are in development stage with some of the vessel fabricators.

11.5 PRESTRESSING OF THICK-WALL VESSELS

It is seen in section 11.2 that the stress in the inner surface of the cylinder is higher than the outer surface. The distribution of the stress can be made more favourable by pre-stressing the inner surface of the cylinder.

Just to show the difference in stress distribution in vessel with and without prestressing, the following two equations are cited:

$$\sigma_\theta = p_i \frac{K^2 + 1}{K^2 - 1}$$ — from Lame's solution (not prestressed)

$$\sigma_\theta = p_i \frac{1}{K^2 - 1}$$ — after prestressing and application of pressure (ideal condition)

The pre-stressing can be done by various methods, such as:

(1) Autofrettage
(2) Shrink fit of cylinders
(3) Ribbon construction of vessel
(4) Mitsubishi coil layer vessel
(5) Multi layer construction.

In all these cases it is possible both to use high yield strength material and to achieve more uniform stress distribution across the thickness.[3]

11.5.1 Autofrettage

Autofrettage is a French word. It means self hooping equivalent to prestressing by means of shrink fitting successive barrels. This is done by applying the hydraulic pressure in thick wall vessel made by forging or from a fabricated thick plate. Since the inner parts of the shell are subjected to high stresses the inner parts of the shell will first exceed the yield point and undergo plastic deformation, whereas the outer parts will still be subjected to elastic deformation.

This produces a greater unit strain in the inner portion of the shell than the outer portion. On release of the overstressing pressure, the difference in unit elongation results

---

[3] Proceedings of Advanced Summer School on Analysis and Design of Pressure Vessels and Piping : M N.R.E. College, Allahabad (India), 1972.

in a residual compressive stress in the inner and a residual tensile stress on the outer portion of the shell.

The thickness of the vessel which undergoes plastic deformation depends on the pressure, what is called auto-frettage pressure. The yielding of inner surface will occur only when the maximum shear stress at that point becomes equal to yield point.

If the pressure is further increased, the plastic deformation patterns will go into the entire wall thickness of the vessel.

The calculation of stress in the plastic region is very difficult as its deformation cannot be predetermined by calculation, it can only be done by extensive experiments.

Moreover, it is very difficult to get a heavy homogeneous thick shell.

For this reason the application of this technique is limited, mostly for gun barrels and small size vessels.

11.5.2 Shrink fitted shell

Application of uniform stress distribution by the method is also very limited. These are normally used for hydraulic cylinders, which are small in size. This method is not applicable for big reactors because of its manufacturing difficulties.

In this process, two or more cylinders are shrunk one over the other. The individual shell thickness range usually between 25 to 75 mm. The inside and outside diameters are predetermined so as to give the required interference fit with outside of the inner cylinder. The outer cylinder is heated to such a temperature that the same can be easily pushed on the inner cylinder which is at room temperature or even cooler and when the outer cylinder cools down it exerts a compressive stress on the inner cylinder.

11.5.3 Ribbon and wire wound vessel

This process is similar to the wire wounding for reinforcing gun barrels. This technique has been extended to include the use of plate and inter locking ribbon for pre-stressing to get uniform stress distribution. The inter locking type strips are applied at fairly high temperature to enable tight holding together of successive layers. The hot application of strip further caused a pre-loading of inner shell and layer. The inner shell is made from about 25 to 40 mm thick plate, depending upon the service, and grooves are machined on the outer surface and then wound by strip specially made in steel mill with close tolerances about 65 to 85 mm wide and around 6 mm thick with projections of 8 mm × 2.5 mm at a pitch about 22 mm.

This type of construction was developed in Germany. Still this type of vessels are in production in U.K., France, Japan and West Germany. The hoop stress in the shell is borne by the tensile force in the ribbon and the axial stress by the shear in interlocks of the ribbons.

11.5.4 Coil layer vessel

This vessel has been developed in Japan with the same concept of average stress distribution. The individual shell ring is made by coiling a continuous strip of light gauge materials 3 to 4 mm thick and about 2 m wide on fabricated inner shell called the inner vessel made from 25 to 32 mm thick plate. The coiling is done under heat and tension which on cooling gives pre-tension on the inner cylinder. The stress distribution and the stress pattern will be almost same as the ribbon construction vessel. The full vessel is made by welding the ring together by circumferential weld. There is no longitudinal seam in the cylinder. The coiling is started by a taper piece and also finished after required thickness is reached by a taper piece.

11.5.5 Multi layer vessel design and construction

In case of shrink fitted vessel construction (section 11.5.2) the necessary interference for prestressing can be obtained with a very small increase in the temperature of the outer ring by amount say 20 to 30°C assuming uniform heating of the ring and no clearance for assembly. It follows that if a narrow longitudinal band of 1/10 to 1/20 of the circumference of the outer shell were heated to 10 to 20 times this temperature difference, the same effect would be produced by cooling. In the cooling of a longitudinal welded joint such an effect can be produced. This provides, therefore, a possible convenient method of prestressing the shells. The Bharat Heavy Plates and Vessels Ltd., follows the same principle of weld shrinkage for construction of high pressure vessel using comparatively thinner plates layer after layer until the design thickness is achieved.

11.5.5.1 *Advantage of multi layer construction*

If required, the inner shell can be made out of corrosion resistance materials with low cost. The 6 mm layer material gives better mechanical properties due to its uniformity throughout. It gives better notch toughness properties than solid wall vessel. In case of any failure, this vessel fails with ductile fracture (i.e., it deforms before ultimate failure) not with brittle fracture ensuring better safety of personnel. The facilities for field assembly without stress relieving reduce trouble for large shipment. Because of thin plate, it does not call for post weld heat treatments, and also fabrication cost is comparatively less.

But due to low thermal conductivity the multilayer vessel will be subjected to more thermal stresses than the solid wall vessel.

Design Example 11.1 : A vessel is to be designed to withstand an internal pressure of 150 MN/m². An internal diameter of 300 mm is specified, and a steel having a yield point of 450 MN/m² has been selected. Calculate the wall thickness required by the various theories with a factor of safety, 1.5.

Solution :
(a) Maximum principal stress theory :
By Eq. 11.3.1
$$\sigma_y/1.5 = p_i (K^2 + 1)/(K^2 - 1)$$

or, $\dfrac{K^2 + 1}{K^2 - 1} = \dfrac{\sigma_y}{1.5\, p_i} = \dfrac{450}{(1.5)(150)} = 2$

or, $K = 1.74$

so, $D_o = (1.74)(300)$
$= 522$ mm

and $t = 111$ mm

(b) Maximum strain theory :

By Eq. 11.3.2

$$\dfrac{\sigma_y}{1.5} = p_i \left[\dfrac{(1 - \mu) + (1 + \mu) K^2}{(K^2 - 1)} \right]$$

or, $\dfrac{0.7 + 1.3\, K^2}{(K^2 - 1)} = 2$

or, $K = 1.96$

so, $D_o = (1.96)(300)$
$= 588$ mm

and $t = 144$ mm

(c) Maximum strain energy theory :

By Eq. 11.3.3

$$\dfrac{\sigma_y}{1.5} = \dfrac{p_i \sqrt{6 + 10\, K^4}}{2\, (K^2 - 1)}$$

or, $\dfrac{6 + 10\, K^4}{(K^2 - 1)^2} = 16$

or, $K = 2.24$

so, $D_o = (2.24)(300)$
$= 672$ mm

and $t = 186$ mm

(d) Maximum shear theory :

(i) By Eq. 11.3.4

$$\dfrac{\sigma_y}{1.5} = \dfrac{2K^2}{(K^2 - 1)}\, p_i$$

or, $\dfrac{K^2}{K^2 - 1} = 1$

It does not give a solution. This theory is not applicable for $\dfrac{\sigma_y}{1.5\, p_i} \leqslant 2$.

In all the solutions 1.5 is used for factor of safety.

(ii) By Eq. 11.3.5

$$\frac{\sigma_y}{1.5} = \frac{\sqrt{3}\, K^2}{K^2-1} p_i$$

or, $\quad \dfrac{K^2}{K^2-1} = 1.16$

or, $\quad K = 2.7$

so, $\quad D_o = (2.7)(300)$
$\quad\quad\quad = 810$ mm

and $\quad t = 255$ mm

Discussion: The required thickness of the vessel shell under consideration according to the equations for the four theories are as follows:

Theory	Thickness (mm)
(a) Maximum principal stress	111
(b) Maximum strain	144
(c) Maximum strain energy	186
(d) Maximum shear	255

It is clear that the various theories give widely different values for the required thickness. Depending on the material, experimental results agree with one theory or other. It is concluded, for example, that steels with high tensile strength (present example) would be expected to fail by shear.[2]

In the case of a vessel of monoblock construction the term $\dfrac{K^2}{K^2-1}$ in Eq. 11.3.5 approaches unity at high values of K, and the maximum internal pressure, p_i, becomes equal to $\sigma_y/[(\sqrt{3})(F)]$ as a limiting condition, where F is the factor of safety. For the present example, this limiting pressure would be, from Eq. 11.3.6, 175 MN/m². This is a serious restriction for monoblock construction, but may be circumvented by other procedures such as prestressing and multilayer construction.

Design Example 11.2: Monoblock vessel construction has many limitations. Prestressed multilayer vessel is also having constructional difficulties. For the above example, will it be possible to design a jacketed or multi-jacketed high pressure vessel eliminating the difficulties of both the construction?

Solution: Monoblock construction has two major limitations. These are plate thickness and limiting pressure. In multilayer vessel construction the major difficulty is in achieving predetermined prestressed condition. In the process of prestressing, a residual compressive stress is developed at the inner surface of the shell. When such vessel is put under internal pressure, a favourable stress distribution occurs in the shell.

The idea of jacketed vessel construction is to provide an external pressure simultaneously while under operation. This will minimize the effect of high internal pressure on the shell.

In this problem let it be assumed that a pressure $p_o = 50$ MN/m² is acting inside jacket, while the inside shell pressure is 150 MN/m².

By Eq. 11.2.3

$$\sigma_\theta = \frac{p_i D_i^2 - p_o D_o^2}{D_o^2 - D_i^2} + \frac{(p_i - p_o) D_i^2 D_o^2}{D^2 (D_o^2 - D_i^2)}$$

Where,
$\sigma_\theta = \dfrac{\sigma_y}{F} = \dfrac{450}{1.5} = 300$ MN/m²

$p_i = 150$ MN/m²

$p_o = 50$ MN/m²

$D_i = 300$ mm

$D_o = D_i + 2t$

$D = D_i$ for maximum stress

$\dfrac{D_o}{D_i} = K$

Substituting the value,

$$300 = \frac{150 - 50 K^2}{K^2 - 1} + \frac{100 K^2}{K^2 - 1}$$

$$= \frac{150 + 50 K^2}{K^2 - 1}$$

or, $K = 1.34$

so, $D_o = (1.34)(300)$

$= 402$ mm

and $t = 51$ mm

External jacket thickness is calculated by Eq. 11.2.6

$300 = 50 (K^2 + 1)/(K^2 - 1)$

or, $K = 1.18$

so, $D_o' = (1.18)(D_i')$

Where, $D_o' =$ jacket outside diameter

$D_i' =$ jacket inner diameter

$= 410$ mm (assumed)

Hence, $D_o' = (1.18)(410)$
 $= 484$ mm
and $t' = 37$ mm

Discussion: From the solution of previous example a monoblock construction requires a shell thickness of 111 mm according to maximum principal stress theory (Lame's solution). If a jacketed construction is made, two shells of 51 mm and 37 mm thick are required. Fabrication of thinner shell is easier. Metallurgical properties of thinner shell are expected to be more uniform. Material of construction for inner shell and jacket may be different. But construction of jacketed high pressure vessel may pose some other problems like connection of jacket with the shell, end connections, nozzle connections, maintaining and controlling pressure, etc. These problems, however, can be solved.

CHAPTER 12
MATERIAL SPECIFICATIONS

12.1 INTRODUCTION

In designing any process vessel, material of construction is to be specified first. This is a prerequisite for stress analysis. Selection of material is influenced by various factors, though mechanical strength and corrosion resistance are two primary factors. To specify any material for process equipment following informations are necessary and very often compromises are to be made among conflicting properties like cost and corrosion resistance or mechanical strength, etc.

(1) **Nature of service conditions**: Type of loading, service temperature, specific nature of fluid handled.

(2) **Material characteristics**: Strength and other mechanical properties like elongation and reduction in area, notch toughness, hardness and resistance to wear, creep and fatigue strength, etc.

(3) **Processing factors**: Effects of fabrication techniques like forming, cutting, etc., heat treatment and weldability.

(4) **Behaviour in the medium**: Resistance to corrosion or other damage in the environment; specific effects on material and identification of the specific material characteristic relevant to the failure contingency.

(5) **Cost**: Balance of cost considerations against service life and hazards of failure.

(6) **Commercial availability**.

12.2 TYPES OF MATERIALS AND THEIR BASIC CHARACTERISTICS

The materials generally used in process vessel construction may be grouped as follows:

(1) **Steel**: Carbon, low alloy, high alloy and clad with stainless steel or other metals.

(2) **Non-ferrous**: Aluminium, copper, nickel, lead and their alloys.

(3) **Special purpose metals**: Titanium, zirconium, etc.

(4) **Non-metallic**: Plastic, concrete.

12.2.1 Carbon steels

This group of steel is comparatively cheap and most commonly used for process vessels and piping. They are low-carbon steels generally containing less than 0.25 per cent carbon with some Mn and Si. Carbon steel with 0.4 per cent carbon is used for bolts, studs and nuts.

Depending on the degree of de-oxidation, a steel may be rimmed, semi-killed or killed. Rimmed steels are seldom used in process vessel construction, due to their lack of chemical homogeneity. Semi-killed steels correspond to an intermediate de-oxidation state between rimming and killing. They are the cheapest steels that may be used for conventional, light duty services. Almost all the plates used in process vessels upto 25 mm thick for light duty service is semi-killed. Fully de-oxidised, silicon killed steels are more homogeneous. They are more expensive and are used for thicker vessels, or, for more severe duty. Al is frequently added in small quantity as grain refiner, to improve the notch toughness of the material.

Carbon steels are generally used in the normalized condition, i e., after a heat treatment consisting of heating over the upper critical point (850—900 °C) and cooling down in air.

Carbon steels have low or intermediate strength at room temperature, with good ductility and are easily formed and welded. Their tensile strength at high temperature is reduced and temperatures of over 450—500 °C produce rapid loss of metal due to scaling. They are suitable for non-corrosive services at ordinary or moderately elevated temperatures.

12.2.2 Low alloy steels

This group consists of steels with alloying elements less than 10 per cent. The alloying elements generally used are Cr, Mo, Ni, and contribute to increased strength, and improved resistance to scaling, oxidation and graphitization at elevated temperatures. They also have higher creep-rupture strength than ordinary carbon steels. These steels require greater control of processing during fabrication and welding. In general Cr and Mo improve mechanical properties, especially at high temperatures, and these steels find extensive application in elevated temperature service. On the other hand Ni increases notch toughness property of low temperatures (upto −80 °C). The low-alloy Ni - steels, containing 2 - 3.5 per cent Ni, are, therefore, used where resistance to brittle fracture at low temperature is important. Though more expensive than ordinary C - steel, their improved properties permit the use of higher design stress and result in a considerable saving in material.

12.2.3 High alloy steels

High alloy steels are used in power industries, and in nuclear and chemical installations, when it is necessary to have corrosion resistant material and whenever contamination has to be avoided.

High alloy steels are of two types :
(1) Ferritic stainless steels
(2) Austenitic stainless steels

The first kind contains only Cr as alloying element and its content ranges from 11-25 per cent. These have excellent resistance to corrosion and oxidation at high temperatures. The austenitic stainless steels, however, constitute the largest group of corrosion resistant steels for a variety of applications in the chemical industry. These are Cr - Ni alloy steels having an austenitic structure and are non-magnetic. Cr contents vary between 18 and 25 per cent and Ni contents between 8 and 20 per cent. The C content varies from 0.04 to 0.25 per cent. They have good strength and ductility. They are not hardenable by heat treatment. But they have a high strain hardening coefficient. That means, their tensile strength and hardness increase appreciably on cold working. They are readily formable and easily weldable, though special precautions are necessary during fabrication to preserve their optimum corrosion resistance. These steels are suitable for high temperature service.

12.2.4 Clad steels

A special product combining corrosion resistance of austenitic or ferritic stainless steels, nickel or nickel alloys with the economy of carbon steel is clad steel. This is a composite with a base material of carbon steel and cladding of stainless steel on the surface, 2–5 mm thick, which is homogeneously bonded to the backing material by rolling.

The mechanical properties of the clad steel plate are the same as those of the backing steel. At the same time, the corrosion resistance is equal to that of the cladding material, provided that adequate care is taken in obtaining a homogeneous and sound cladding of a sufficient thickness, usually 10–20 per cent of the total thickness of 2–5 mm.

Cladding by weld deposition is the only practicable method for forged components such as nozzles and flanges. It is also the only method that ensures good adherence between cladding and backing plates, when the total thickness required is above 50 mm. In this case, the weld deposition may be made before or after forming the plate into the required shape. After the deposition of cladding, it is ground or machined to the specific surface finish.

12.2.5 Non-ferrous metals

Non-ferrous metals are frequently used for special applications in the chemical industry, the most common materials being aluminium, copper, nickel and their alloys. Platinum, silver and lead are used either alone or in the form of liners. In addition, titanium, zirconium, niobium and other find increasing applications.

12.2.5.1 *Aluminium and aluminium alloys*

Aluminium and its alloys are readily available and relatively inexpensive. These are very light and are non-toxic, non-magnetic and good reflectors of radiant heat. They have a high thermal conductivity. Pure aluminium has high resistance to corrosion in chemical environments because of the formation of their protective film of oxide on its surface on exposure to air. Pure aluminium has low strength and is soft and ductile.

All aluminium alloys retain their ductility at sub-zero temperatures. They are therefore of special interest for the construction of low temperature process vessels, operating at temperatures of the order of — 200 °C. These are widely used in dairy and food processing and refrigeration industry.

Besides this application for the construction of low temperature vessels, aluminium and its alloys are used for handling most organic solvents, food stuffs, hydrogen peroxide, nitric acid, sea water and inorganic salts. Fresh water may cause some pitting. Due to the reduction of mechanical strength with temperature, this material is seldom used for vessels operating above 150 °C.

12.2.5.2 *Copper and copper alloys*

Copper and its alloys are used in food processing plants, in the manufacture and recovery of organic solvents and in heat exchangers and evaporators for general purposes.

The use of copper is limited to fairly low temperature applications, due to its poor creep strength above 150 °C.

Copper has good resistance to corrosion in the presence of sea water and a large number of chemicals. It should not be used in contact with oxidising acids, ammonia and carbon dioxide solutions, metallic salts susceptible to reduction and turbulent sea water.

Brasses are copper-zinc alloys, with a zinc content as high as 45%. Their corrosion r sistance is not as high as copper, especially when the Zn - content is higher than 20 per cent due to the loss of Zn. These high - zinc brasses may fail by stress corrosion cracking. Their application is limited in general to tubes and tube plates for heat exchangers, operating at low temperatures.

Tin bronzes are mainly used for castings, such as, valve bodies. Other bronzes of wide applications are Al - Cu alloys. These are very resistant to oxidation, corrosion and stronger than pure copper, especially at temperatures above 150 °C. The 10 per cent Al - bronze, with Fe and Ni is used for tubes and tube plates, upto 350 °C. All Al - bronzes resist mineral acids and have excellent resistance to sea water, even under turbulent conditions.

Cu-Ni alloys are widely used in steam condensers, operating with polluted cooling water or sea water.

12.2.5.3 *Nickel and nickel alloys*

Nickel and its alloys are used for handling a variety of corrosive fluids at low or elevated temperatures and where it is essential to prevent the contamination of contained fluid. Nickel and its alloys are very expensive and it will be found that in process vessel construction they are used only as cladding materials. The only excepts are heat exchanger tubes and tube plates, and relatively small, thin-walled vessels.

Following nickel alloys are widely used :

(1) Monel : A Ni-Cu alloy containing 67 per cent Ni and 30 per cent Cu combines high strength, ductility and excellent corrosion resistance in a number of chemical media. Widely used in marine water service and in organic chemical industry. Used also as lining or cladding material on carbon steel or low-alloy steel.

(2) Inconel : A Ni-Cr-Fe alloy containing 16 per cent Cr, 8 per cent Fe, 1 per cent Mn and balance Ni has good strength and toughness at both elevated and low temperatures, with excellent corrosion resistance. Widely used in dairy and food processing equipment, for alkalies and alkaline solutions, organic acids specially useful for high-temperature applications, where resistance to oxidation, nitriding or halogen-bearing atmospheres is important.

(3) Incoloy : Another Ni-Cr-Fe alloy containing 21 per cent Cr, 44 per cent Fe, 1.5 per cent Mn and balance Ni. Its properties are similar to that of Inconel. Essentially it is a high temperature alloy, with high creep-rupture strength and structural stability at elevated temperatures of 850 - 900 °C.

12 3 MATERIAL SPECIFICATIONS FOR SPECIFIC ENVIRONMENTS

12.3.1 Elevated temperature services

For high temperature service the important material properties that need be considered are short-time tensile properties, creep strength at high temperature, internal and surface stability (resistance to scaling and corrosion). These are all affected by composition, microstructure and grain size. Plain carbon steel has shortcomings in relation to all the above properties at high temperature. It has low creep strength, poor internal stability (tends to decarburization and graphitization at temperatures of 400 °C and over), poor surface stability (susceptible to excessive scaling at temperatures of 500 °C and over).

The addition of Cr, Mo and V improves strength at elevated temperatures as well as resistance to scaling and oxidation. The higher the Cr - content the greater is the limiting temperature without excessive scaling. Approximate limiting temperatures for various steels are shown as below :

C - steel	:	400 °C
C - $\frac{1}{2}$ Mo steel	:	500 °C
$1\frac{1}{4}$ Cr - $\frac{1}{2}$ Mo steel	:	540 °C

2¼ Cr - 1 Mo	:	560 °C
5 Cr - ½ Mo	:	620 °C
12 Cr	:	700 °C
18 Cr - 8 Ni	:	900 °C
27 Cr	:	1 100 °C
25 Cr - 12 Ni	:	1 040 °C
15 Cr - 20 Ni	:	1 100 °C

At high temperature service one should note that the presence of some particular bases in the environment affects the behaviour of the steel. For example, H_2 in the environment can cause hydrogen embrittlement at high pressures and temperature. For hydrogen service austenitic stainless steels are found to be quite suitable.

Similarly, gaseous environments containing S and S - compounds are corrosive to ordinary steels at high temperatures. They also attack austenitic stainless steels. In such environments, which are commonly met with, in refineries processing sour (S - containing) crudes, 5 Cr - ½ Mo steel provides good corrosion resistance in addition to high temperature strength. For more severe conditions, 9 Cr - 1 Mo steel provides better protection.

For heater tubes, furnace tubes and other furnace parts, which need both creep-rupture strength and internal stability and resistance to scaling, the 25 Cr - 12 Ni and 25 Cr - 20 Ni austenitic stainless steels offer the best combination for temperatures of 850 °C and above. Reformer tubes for steam-naphtha reforming are made of centrifugally cast high C - 25 Cr - 20 Ni stainless steel and headers, etc., made of Incoloy.

½ Mo steel which was originally used for high temperature steam plants, has also been found suitable in petroleum cracking-furnaces, at temperatures between 500 and 550 °C. But due to certain reported failures, it is now being replaced for high temperature service by Cr - Mo steels, with Cr - contents of ½ - 5 per cent and Mo - contents of ½ - 1 per cent.

½, 1, 2¼, 3 and 5 per cent Cr have better corrosion resistance than ordinary C-steel and they are used for high temperature service under mild corrosive conditions. They are of special interest in coal hydrogenation processes and in the synthesis of anhydrous ammonia. These processes involve pressures of the order of 65 MN/m^2 and temperatures as high as 500 °C.

1 Cr - ½ Mo steel is used for pressure vessels, steam and super-heater tubes and piping systems designed to operate at temperatures below 500 °C.

Heavy duty bolting material is generally medium C - 1 Cr - Mo steel with or without vanadium. The maximum service temperature suggested is 500 °C. For temperatures above 450 °C, the use of 1 Cr - Mo - V bolts are desirable, in spite of the very poor notch toughness of this steel.

In petroleum industry, 18 Cr - 8 Ni stainless steels are usually specified as having good corrosive resistance upto about 800 °C. Above this temperature, upto about 1 100 °C, 25 Cr - 12 Ni or 25 Cr - 20 Ni stainless steels are generally selected. In addition, several types of austenitic Cr - Ni - Mo steels, with V and other alloying elements have been developed for high temperature service due to their good stability, corrosion resistance and creep strength.

Ferritic steels are used in chemical plant, usually for non-pressurized components.

12.3.2 Low temperature services

Many chemical processes are operated at sub-zero temperatures. In specifying metals and alloys for low temperature services (below 0 °C), the most important mechanical property is notch-toughness, that means, the ability to deform plastically at stress-concentrations and resistance to brittle fracture in conditions of multi-axial stresses, at notches.

Ferritic steels exhibit sharp decrease in notch-toughness at low temperatures, as measured by charpy impact tests, not evident in the oridinary tensile test. A large group of non-ferrous alloys and austenitic stainless steels show a good correlation in ductility in the behaviour of notched specimens and in the ordinary tensile test, even at very low temperatures.

In the design of process vessels and piping, fracture-safe design is essentially based on selection and testing of material to prevent brittle crack initiation. The notch toughness of steels is influenced by a number of factors, such as, chemical composition, deoxidation and melting practice, fabrication and heat treatment, section thickness. The effects of these factors are discussed below :

(1) Chemical composition : An increase in C - content increases the transition temperature below which metals become brittle. An increase of 0.1 per cent can raise transition temperature by as much as 15 - 20 °C. Mn helps in lowering the transition temperature, so that Mn/C ratio is a more significant parameter and the lowering of this ratio tends to raise the transition temperature. Alloying with Nb reduces transition temperature.

(2) Deoxidation and grain size : Fully killed steels, refined by addition of Al or Nb have a fine grained structure resulting in lower transition temperature. Normalizing has a similar beneficial effect.

(3) Thickness : The thickness of the section has an important effect on its behaviour. An increase in thickness by 25 mm can raise transition temperature by as much as 10 - 15 °C, because increase in thickness produces transverse constraint and inhibits plastic deformation.

(4) Cold working : Cold working in forming operations to the extent of 10 per cent or more extreme fibre elongation can raise transition temperature by 10 - 15 °C.

(5) Residual stresses : The presence of residual stresses as in welded structures promotes brittle behaviour. Relief of these residual stresses by post weld heat-treatment has a favourable effect.

Following are the examples of a few metals and the temperature range :

Silicon killed, boiler quality C-steel can be used in low temperature vessels down to about $-60\,°C$, in thickness below 12 mm in stress relieved condition.

The $3\frac{1}{2}$ per cent Ni - steel has been developed for the fabrication of process vessels for low temperature service (not below $-100\,°C$) A 9 per cent Ni - steel is used for similar service at temperature not lower than $-190\,°C$.

In general the austenitic stabilized steel types 18 Cr - 8 Ni - low C are used in welded construction for relatively low temperature service. Table 12.1 specifies steels for low temperature service.

Table 12.1 Steels for Low Temperature Service[1]

Temperature Limit (°C)	Service	Boiling Point at Atm (°C)	Type of Steel	Steel Specification
-50	Sulphur dioxide	-10.0	Carbon steel	IS : 2002 Gr 1
	Methyl chloride	-23.7		IS : 1570 $-$ 11 Mn 2
	Freon 12	-30.0	Low Cr. steel	
	Ammonia	-33.3		
-51 to -80	Hydrogen sulphide	-60.0	Low alloy steel	IS : 2100 Gr 1
	Carbon dioxide	-78.0	2.5 per cent Ni steel	
-81 to -100	Acetylene	-84.0		
	Ethane	-83.3	3.5 percent Ni steel	
	Nitrous oxide	-89.5		
-101 to -200	Ethylene	-104.0	9 percent Ni steel	IS : 1570 04 $-$ Cr 19 Ni 9
	Methane	-161.4	Austenitic	
	Oxygen	-183.0	Stainless steel	IS : 1570 $-$ 07 $-$ Cr 19 Ni 9 Mo 2 $-$ Ti 28
	Argon	-186.0		
	Fluorine	-187.0		
	Nitrogen	-195.8		
-201 to -270	Neon	-246.0	Austenitic stainless steel	
	Hydrogen	-252.7		
	Helium	-269.0		

[1] Proceedings of Advanced Summer School on "Analysis and Design of Pressure Vessels and Piping", MNREC, Allahabad, 1972.

In short the nature of brittle fracture is discussed below :

Brittle fracture is a sudden catastrophic fracture, which occur without prior indication of general deformation at nominal stresses below yield point. However, it can occur only under the following conditions :

(1) The presence of a crack-like defect of sufficient size and depth.

(2) The presence of local stresses near yield in the region of the defect.

(3) A temperature low enough for brittle crack propagation in the material.

12 3.3 Corrosive services

As a threat to plant equipment, personnel safety and product purity, corrosion is a matter of vital concern to every chemical engineer. Design engineer tries to select the most practical material of construction for each case of corrosive service by proper balance of corrosion resistance, mechanical properties and cost. Each of these is equally important in making the final choice. Under mechanical properties he considers structural characteristics and methods of fabrication. Corrosion often forces the design engineer to apply an uncertainty factor where corrosion rates are unknown. This leads to the wastage of metal through equipment overdesign.

Corrosion processes involving metals can be classified into two types : Oxidative and non-oxidative. The former is by far the most common type since it includes both ordinary atmospheric corrosion and chemical attack by acids and corrosive gases.

In oxidative corrosion, the metal undergoes chemical oxidation, losing electrons and increasing in positive valence - essentially reverting to its ore, a more stable form. The laws of nature, of course, require that another substance - usually the environment or corroding agent - be simultaneously reduced.

It has been established that at ordinary temperatures, this oxidation-reduction takes place by an electrochemical mechanism through the formation of spontaneous galvanic cells in the presence of moisture. Only at very high temperatures (above 800 °C) will dry gases directly attack and oxidize most metals in their usual structural form.

Non-oxidative corrosion occurs in handling liquid metals and presents serious problems because it is not yet as well understood as the chemical type. In non-oxidative corrosion, a metal dissolves or disintegrates but is not chemically oxidized.

The rate of oxidative corrosion is influenced by a number of modifying factors like light, heat, surface homogeneity and the presence or absence of natural or artificial polarizers.

Variations in these factors produce different patterns of attack which are classified under eight general categories :

(1) Uniform corrosion.

(2) Pitting.

(3) Bi-metallic corrosion.

(4) Concentration cell corrosion.

(5) Intergranular corrosion.

(6) Stress corrosion.

(7) Dezincification.

(8) Erosion-corrosion.

Uniform corrosion is the most common type of attack and accounts for most of the annual losses of metal on a weight basis. It is also the least intense in its action and is relatively easy to predict and control. As a safeguard against such corrosion extra thickness as corrosion allowance based on decay rate per year and expected service life is provided.

The other more localized and intense forms of attack often cause more serious failures. For example, pitting corrosion occurs at small local areas without general corrosion and results in pitting which may lead to decaying of the metal wall. This is caused due to the formation of local galvanic couples oxidizing the metal. Among the austenitic stainless steels, the addition of 2-3 per cent Mo makes pitting less severe.

As the name implies, intergranular corrosion is localized attack at the grain boundaries of a metal or alloy causing cracking and failure without general loss of metal. Unstabilized stainless steels are especially susceptible to intergranular corrosion because of chromium carbide precipitation at the grain boundaries. This happens generally in the temperature range of 400 - 900 °C when structural change in austenitic stainless steel occurs. In this condition the steel is said to be 'sensitised' and become easy prey to intergranular attack by a number of corrodents, which otherwise may cause little general corrosion. (Examples are HNO_3, H_3PO_4, etc.) In this type of attack, complete failure of the metal results even though only a small fraction of the total weight corrodes. Lowering of C-content to less than 0.03 per cent maximum or addition of stabilizing agent like Ti or Nb offer protection by eliminating the precipitation of chromium carbides. In the ordinary grades, heat treatment by annealing at 1 000 - 1 000 °C, followed by rapid cooling also provides protection.

Stress corrosion is the general term applied to several patterns of attack in which stress is believed to accelerate corrosion. Stress corrosion includes simple stress corrosion, stress-corrosion cracking and corrosion fatigue. Stress corrosion often occurs when uniform corrosion is nearly negligible. To produce cracking, stress corrosion requires the presence of rather severe residual tensile stresses such as those resulting from cold working or improper welding. Austenitic stainless steels are susceptible to stress corrosion cracking in hot chloride solutions, carbon steel in hot caustic solutions and hot nitrates, brass in ammonia environments. Since the attack is specific to a metal in a particular corrodent, a solution can be found by choice of different metal (usually high-alloy and hence costly) or by suitable heat treatment for relief of residual stresses.

MATERIAL SPECIFICATIONS

With the present knowledge of corrosion mechanism, it is not possible to predict how long a susceptible material will survive in a given environment, since the interrelations between temperature, corrosion environment, stress level and alloy compositions have not yet been clearly defined.

The most acceptable current theory is that cracking results from electrochemical action between narrow paths of stress-caused anodic material and the more cathodic bulk surface. As corrosion proceeds, the tensile stresses open crevices exposing more anodic metal to attack.

Well-known examples of stress corrosion are caustic embrittlement of iron and steel and season cracking of brass. Corrosion fatigue occurs when a metal is subjected to repeated reversals of stress in a corrosive medium. These stress reversals lend to uneven surface depressions or pits which eventually reduce the metal cross-section to the point of tensile failure.

Dezincification failure was originally observed on brasses from which Zn was selectively leached by corrosive environments. Graphitization of cast iron is also in this category. Here iron is corroded away, leaving a weak graphite network.

Increasing the flow rate of a corrosive liquid past a metal surface generally increases corrosion rate. Corrosive products and protective films are swept away. At high velocities a type o attack called erosion-corrosion results. It is a combination of mechanical wear and corrosion and is characterized by smoothly formed interconnected pits. The presence of abrasive materials in the moving fluid intensifies this attack. Pipe, fittings, agitators and pumps are examples of process equipment which are affected by erosion-corrosion. Cavitation-erosion, a special case of erosion-corrosion, occurs in highly turbulent flow where vapour cavities are formed in the liquid.

Therefore, the corrosion of metals and alloys in various chemical environments depends on many factors — concentration of corrodant, temperature, presence of impurities presence of O_2 whether stagnant or flowing, etc. Considerable data are available from extensive tests carried out in the laboratory and in plant corrosion tests. Ultimately, choice will be based on prior experience in particular environment and a balance of cost and service life desired.

Materials used for some of the more common corrosive services are given below :

(1) For acid salts solutions in most cases - IS : 1570-1961 (Cr 19 Ni 9 and Cr 18 Ni 11 Mo 3).

(2) HCl - air free, concentration upto 20 per cent at room temperature - Monel.

(3) HF - all concentrations upto 92 per cent and temperature 110 °C - Monel (austenitic stainless steel is subjected to severe attack.)

(4) H_2SO_4 - over 98 per cent and cold - ordinary carbon steel (also for sulphurous acid), but lower than 20 per cent and over 85 per cent - IS : 1570-1961 (Cr 18 Ni 11 Mo 3) and for all other concentrations, where IS : 1570-1961 is attacked - Cr 20 Ni 25 Mo 3 with 2-3 per cent Cu or similar copper bearing stainless steel. Also lead and lead lining.

(5) HNO_3 - All concentrations upto boiling point - IS : 1570-1961 (Cr 19 Ni 9 and Cr 19 Ni 9 Nb) ; temperature above boiling point - Cr 25 Ni 12 Nb and Cr 25 Ni 20 Nb.

(6) H_3PO_4 - for most concentrations - IS : 1570-1961 (Cr 18 Ni 11 Mo 3) ; for more severe conditions - a copper bearing stainless steel with Cr 20 Ni 29 Mo 2.5 Cu 3.5 stabilized with Nb ; also rubber lining.

(7) Carbonic acid - IS : 1570-1961 (Cr 19 Ni 9).

(8) Urea and carbamate - IS : 1570-1961 (Cr 18 Ni 11 Mo 3).

(9) Marine waters - Carbon steel, Al-bronze, cupronickels, Monel 400.

During the past two decades, reinforced plastics have become established as a regular material of construction. Ti and a number of its alloys have become commercially available at reasonable cost. Polyesters and phenolics show excellent heat and corrosion resistance. Polyethylene and its fluoronated derivates are chemically inert and tough. Though costlier than steel, for many applications this may be offset by lower fabrication expenses, lower installation cost and higher strength-to-weight ratio.

Titanium possesses a number of remarkable corrosion resistant properties. But its resistance is poor in hot or concentrated reducing acids, hot concentrated Al - or Zn - chlorides, dry chlorine and ionizable fluorides.

12.4 INDIAN STANDARDS ON MATERIALS

Following list with title gives a few selective numbers of standard related to materials of construction for process vessels. The mechanical properties of these materials will be available in the Appendix.

No. of the Standard	Title
IS : 407 — 1966	Specification for brass tubes for general purposes.
IS : 410 — 1959	Specification for rolled brass plate, sheet, strip and foil.
IS : 737 — 1965	Specification for wrought aluminium and Al - alloys, sheet and strip.
IS : 738 — 1966	Specification for wrought aluminium and Al - alloys, drawn tube.
IS : 961 — 1962	Specification for structural steel (high tensile).

MATERIAL SPECIFICATIONS

No. of the Standard	Title
IS : 1385 — 1959	Specification for phosphor bronze rods and bars, sheet and strip, and wire.
IS : 1545 — 1960	Specification for solid drawn copper alloy tubes.
IS : 1570 — 1961	Schedules for wrought steels for general Engineering purposes.
IS : 1914 — 1961	Specification for carbon steel boiler tubes and superheater tubes.
IS : 1972 — 1961	Specification for copper plate, sheet and strip for industrial purposes.
IS : 1978 — 1961	Specification for line pipe.
IS : 1979 — 1961	Specification for high test line pipe.
IS : 2002 — 1962	Specification for steel plates for boilers.
IS : 2004 — 1962	Specification for carbon steel forgings for general engineering purposes.
IS : 2040 — 1962	Specification for steel bars for stays.
IS : 2041 — 1962	Specification for steel plates for pressure vessels.
IS : 2062 — 1962	Specification for structural steel (fusion welding quality).
IS : 2371 — 1963	Specification for solid drawn copper alloy tubes for condensers, evaporators, heaters and coolers using saline and hard water.
IS : 2501 — 1963	Specification for copper tubes for general engineering purposes.
IS : 2611 — 1964	Specification for carbon chromium molybdenum steel forgings for high temperature service.
IS : 2856 — 1964	Specification for carbon steel castings suitable for high temperature service (fusion welding quality).
IS : 3038 — 1965	Specification for alloy steel castings for pressure containing parts suitable for high temperature.
IS : 3444 — 1966	Specification for corrosion resistant steel castings.
IS : 3503 — 1966	Specification for steel for marine boilers, pressure vessels and welded machinery structures.
IS : 3609 — 1966	Specification for chrome-molybdenum steel, seamless, boiler and superheater tubes.

12.5 SOURCES FOR INFORMATION

Following are the few valuable references which contain detailed informations regarding material specifications :

1. J. H. Perry, "Chemical Engineers Hand Book", McGraw-Hill Book Co., New York.
2. C. L. Mantell, "Engineering Materials Hand Book", McGraw-Hill Book Co., New York.
3. H. H. Uhlig. "Corrosion Hand Book", John Wiley and Sons, Inc., New York.
4. Proceedings of Advanced Summer School on "Analysis and Design of Pressure Vessels and Piping", MNREC, Allahabad (India), 1972.
5. M. B. Bickell and C. Ruiz, "Pressure Vessel Design and Analysis", Macmillan Co., London.
6. R. V. Jelinek, "Corrosion", Chemical Engineering, 1958.
7. IS : 2825 — 1969 Code for unfired pressure vessels, Indian Standards Institution, New Delhi.
8. L. E. Brownel, and E. H. Young, "Process Equipment Design", John Wiley and Sons, Inc., New York.

CHAPTER 13

EQUIPMENT FABRICATION AND TESTING

13.1 INTRODUCTION

In earlier chapters the design philosophy of process equipment are discussed. With that it is also necessary to know something about equipment fabrication and testing. Fabrication involves various steps like forming, machining, assembling, welding, riveting, etc. Now-a-days riveting is not generally preferred for joining two parts, if the parts are weldable. In this chapter, therefore, welding design will be primarily considered. Testing methods of fabricated equipment are also to be discussed in short. In design and fabrication of equipment certain precautionary steps like avoiding sharp corners, stress raisers, etc. are necessary to minimize corrosion.

13.2 DESIGN OF WELDED JOINTS

In designing welded joints following points are to be noted:

1. Adequate provision of good accessibility for welding. Lack of this will cause inconvenience to the workers reducing their efficiency and increasing costs of welding. Similarly for inspection also good accessibility is to be provided.

2. The optimum design for welded construction is the one with the minimum amount of welding. As large number of unnecessary welds causes distortion and residual stresses, these are required to be avoided. Welds should not be designed to converge in one point. For example, in a cylindrical vessel composed of sections with longitudinal seams, these should be offset.

3. Cracking during welding may occur due to differential expansion. The design of the joint must incorporate sufficient flexibility to prevent this happening.

4. In thick plates, distortion may be controlled by the use of double-V or double-U joint preparations.

13.2.1 Types of welded joints and their specifications

Fig. 13.1 shows some typical joint preparations for butt welding of steel plates of equal thickness. This figure and the following comments are mainly based on the recommendations of the Indian Standards Institute. V and double-V preparations are in general preferred to the U and double-U types, due to the ease and economy with which the edges may be prepared. The amount of metal deposited is higher in the former types of preparations than in the latter.

The specifications and applications of the welded joints shown in Fig. 13.1 are given in the following Table.

212   CHEMICAL EQUIPMENT DESIGN—MECHANICAL ASPECTS

(a) SINGLE 'V' WITHOUT BACKING STRIP

(b) SINGLE 'V' WITH BACKING STRIP

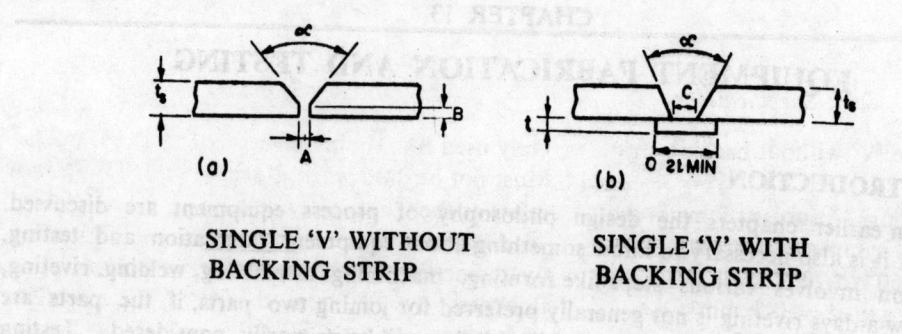

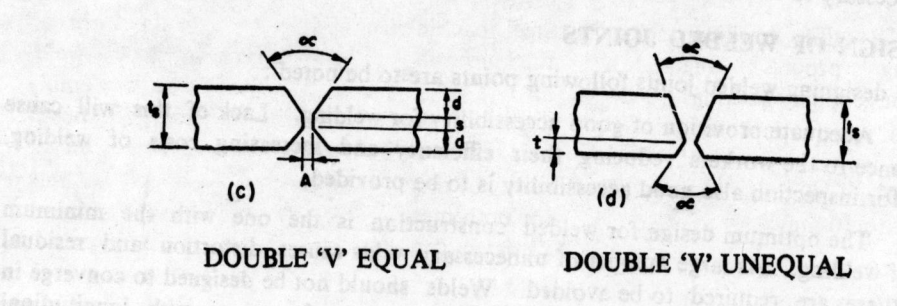

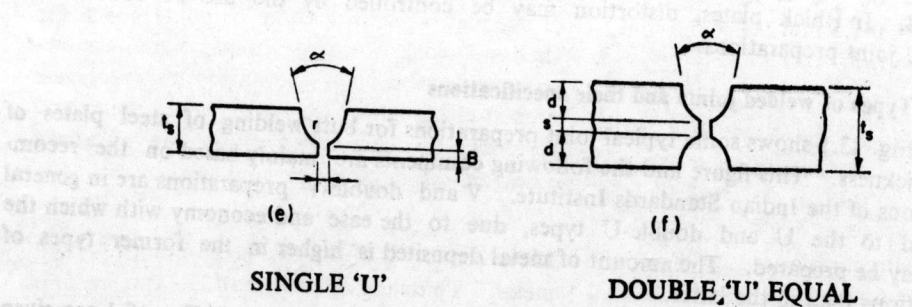

(c) DOUBLE 'V' EQUAL

(d) DOUBLE 'V' UNEQUAL

(e) SINGLE 'U'

(f) DOUBLE 'U' EQUAL

Fig. 13.1 Types of butt welded joints of equal thickness

Table 13.1 Some Details of Butt Welded Joints

Ref. Fig.	Joint Specifications	Applications
(a)	Single 'V' without backing strip. Welding process: metal-arc. Usual symmetrical preparations. $\alpha = 60°$ to $70°$ for all positions. Root gap $A = 0\text{-}3$ mm Root depth $B = 0\text{-}3$ mm	Only used for circumferential joints in cylinders. Must not be used when a tensile bending stress exists at the root or when subjected to cyclic loading, unless special precautions are taken to overcome the possible lack of penetration. For thickness over 20 mm, this preparation may be uneconomical due to the large amount of weld metal required. Precaution against distortion must be taken.
(b)	Single V with backing strip. Welding process: metal arc. (steel or copper backing); submerged arc (copper or bed of flux). $\alpha = 30°$ $C = 8$ to 10 mm	Steel backing for metal-arc processes in all thickness when one side is inaccessible to welding. This type of preparation is preferably limited to thicknesses upto 32 mm. The possibility of stress concentrations and corrosion at the root must be considered. Copper backing for both metal-arc and submerged-arc processes. Greater care needed for using this type of backing to avoid poor penetration.
(c), (d)	Double 'V' Welding process: metal arc and submerged arc $\alpha = 60°$; $\alpha' = 90°$; $S = 0\text{-}1.5$ mm; $A = 0.3$ mm; $t = t_s/3$.	Most commonly used preparation for metal-arc welding of thick section when both sides accessible. Root chipping necessary to achieve sound weld. This may be facilitated by the use of unequal preparations and chipping from the shallow side. Unequal preparations are recommended to obtain the same cross-sectional area at either side of the root after chipping and to reduce the volume to fill in the overhead position. Thickness above 20 mm.
(e), (f)	'U' Preparations. $r = 6$ mm $B = 3$ mm; $S = 3$ mm $\alpha = 20°$ $A = 0\text{-}3$ mm	More economical from the stand-point of weld metal. Thickness greater than 20 mm recommended.

For butt welding plates of unequal thickness a tapered transition with an angle not larger than 14° (1 in 4 slope) should be provided. The weld may be in the transition section, in which case it must be ground flush. Fig. 13.2 shows some typical arrangements. When the weld is not ground flush, it is advisable to leave a distance equal to at least twice the plate thickness between the centre line of the weld and the beginning of the transition.

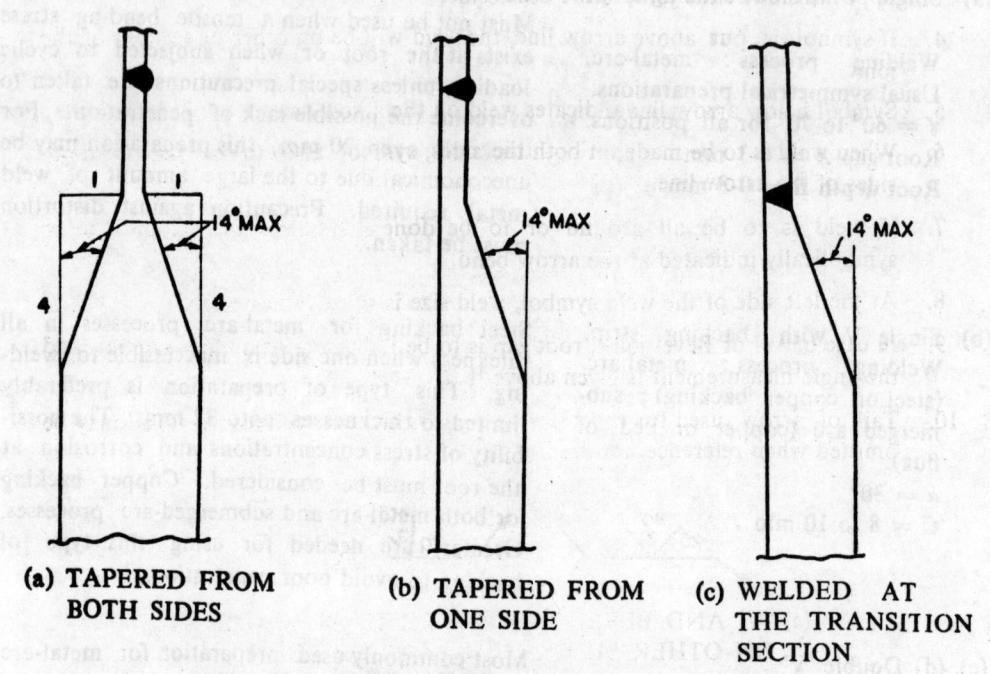

(a) TAPERED FROM BOTH SIDES   **(b) TAPERED FROM ONE SIDE**   **(c) WELDED AT THE TRANSITION SECTION**

Fig. 13.2 Butt welded joints of unequal thickness

Most attachments, such as supporting skirts and brackets, are welded to the vessel by means of fillet welds, without penetration. The design of these welds is given in "Welding Handbook". For the preliminary layout, the approximate dimensioning of the welds may be based on an allowable shear stress equal to 50% of the design stress for the vessel material. Tensile stress may be allowed upto 75% of the same design stress.

13.2.2 Symbolic representation of weldments

In drawings it is not always possible to show the nature of welding in the assembly view. Weldings are indicated by single lines. In that case the type of welding desired by the designer can be indicated by welding symbols as shown in Fig. 13.3. To use the symbols following points are to be remembered[1].

---

[1] L. E. Brownell and E. H. Young, "Process Equipment & Design", John Wiley and Sons, Inc., New York.

EQUIPMENT FABRICATION AND TESTING

1. Symbol of weld is put either below or above the arrow line or on both sides as the case may be.
2. Symbol above arrow line will show the nature of the weld and the same below will be the mirror image.
3. The side of the joint to which the arrow points is the arrow side, and the opposite side of the joint is the other side.
4. If symbol is put above arrow line, the weld will be on other or far side of the joint.
5. Symbol below arrow line indicates weld on the arrow side.
6. When weld is to be made on both the sides, symbol also to be put on both sides of the arrow line.
7. If weld is to be all around or to be done at the field, instruction is to be symbolically indicated at the arrow bend.
8. At the left side of the weld symbol, weld size is to be indicated.
9. In case of 'V' or Bevel joint, root gap is to be indicated inside the symbol and the angle measurement is given above that.
10. Tail of arrow used for specification process or other reference. (Tail may be omitted when reference not used.)

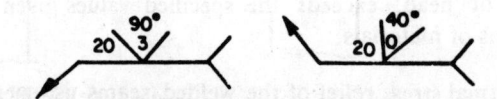

(a) 'V' AND BEVEL JOINTS ARE TO BE MADE ON OTHER SIDE OF THE ARROW POINT

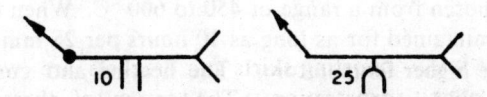

(b) SQUARE AND 'U' WELDS ARE TO BE ON THE ARROW SIDE. COMPLETELY BLACK CIRCLE INDICATES FIELD WELD

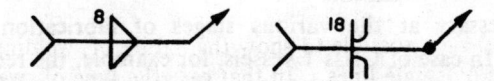

(c) FILLET AND 'J' WELDS ON BOTH SIDES OF JOINT. CIRCLE INDICATES WELD ALL AROUND

Fig. 13.3 Welding symbols for drawing

13.3 Post weld heat treatment

Due to local overheating during welding thermal stresses are induced in the vessel wall. To relieve the residual stress post weld heat treatment is prescribed. In certain cases thermal stress relief is essential ; but in some cases this is not mandatory. In deciding the requirement for post weld stress relief, many factors like material composition, plate thickness, service, etc. are to be considered. For example, vessels or parts of vessels fabricated from carbon and low alloy steels must be thermally stress relieved under following conditions[2].

(a) For containing toxic or inflammable material ;

(b) For operating below 0 °C ;

(c) For use with media liable to cause stress corrosion cracking ;

(d) Subjected to excessive local stress concentration which may give rise to cracking, or subjected to changing loads and subsequent risk of fatigue failures ;

(e) Risk of brittle fracture due to the combined influence of material, transition temperature, notch effect, etc. ;

(f) It is necessary to maintain dimensional accuracy and shape in service ;

(g) The plate thickness including corrosion allowance, at any welded joint in the vessel shell or head, exceeds the specified values given in the standards[2] for various groups of materials.

Procedure for thermal stress relief of the welded seams is specified in the Codes for most vessels and piping. The purpose of this heat treatment primarily, though, to reduce the residual stress build up during welding, depending on the temperature achieved, it may also have a beneficial normalizing effect on the material. The stress relieving temperature is chosen from a range of 450 to 600 °C. When the lower temperature is selected, it has to be maintained for as long as 10 hours per 25 mm thickness, which only 1 hour is necessary for the higher temperature. The heating and cooling rates are given by most Codes like IS : 2825—1969. It is advisable to follow the recommendation of the codes and to check that the temperature stresses during the whole treatment do not exceed the yield stress of the material.

13.4 INSPECTION AND NON-DESTRUCTIVE TESTING OF EQUIPMENT

Inspection is necessary at the various stages of fabrication depending on the classification of vessels[2]. In case of Class 1 vessels, for example, the requirements are :

1. Visual inspection of surface for objectionable defects.
2. Assembly and alignment of vessel sections prior to welding.
3. Identification and stamping of weld test plates.

---

[2] IS : 2825—1969, Indian Standards Institution , New Delhi.

4. Inspection during welding in progress, including second side welding grooves after preparation by chipping, gouging, grinding or machining.
5. Inspection of main seams after dressing.
6. Calibration and dimensional check after completion.

For other two classes of vessels also, more or less, similar inspection schedule is followed.

Depending on the requirement various procedures are recommended by the design codes. A few of them are described in the following sections.

13.4.1 Pressure tests

13.4.1.1 *Hydrostatic pressure tests*

Pressure vessels made of mild or low — alloy steels designed for internal pressure, are subjected to a hydraulic test pressure, which at every point in the vessel is at least equal to 1.3 times the design pressure multiplied by f_1/f_2.

$$\text{Test pressure} = 1.3 \times \text{design pressure} \times \frac{f_1}{f_2}.$$

where, f_1 = allowable stress value of material of construction at test temperature, and

f_2 = allowable stress value at design temperature.

It is important in the interest of safety that the vessel be properly vented so as to prevent formation of air pockets before the test pressure is applied. It is recommended[2] that during the test the temperature of the water should not be below 15 °C.

The vessels are kept under test pressure for sufficiently long time for checking and inspection.

Single wall vessels and chambers of multi-chamber vessels designed for vacuum or partial vacuum only be subject to an internal hydraulic test pressure not less than 1.3 times the difference between normal atmospheric pressure and the minimum design internal absolute pressure, but in no case less than 1.5 bar. In case of jacketed vessels when the inner vessel is designed to operate at atmospheric pressure or under vacuum conditions, the test pressure need only be applied to the jacket space.

13.4.1.2 *Pneumatic pressure tests*

Pressure testing with air or gas may be carried out in lieu of the standard hydraulic test in the following cases :
1. Vessels that are so designed, constructed or supported that they cannot safely be filled with water or liquid ;
2. Vessels that are to be used in such services where even small traces of water cannot be tolerated.

In carrying out pneumatic pressure test adequate precautions, such as blast walls or pits and means for remote observation are essential. The pneumatic test pressure is not less than the design pressure and also not more than test pressure in a hydraulic test.

The procedure of pneumatic pressure test would be as follows. The pressure is gradually to be increased to not more than 50 percent of the test pressure. There after the pressure is to be increased in steps of approximately 10 percent of the test pressure till the required test pressure is reached. Then the pressure is to be reduced to the value of the equivalent design pressure and held at this pressure for a sufficient time to permit inspection of the vessel.

13.4.2 Radiography (X-rays) tests

Radiography is the most certain way of detecting the defects in the welded seams. As it is the costly test procedure, 'Radiography A' covering the radiographic examination of all longitudinal and circumferential butt welds in drums, shells and headers throughout their whole length including points of intersection with other joints is recommended for special cases[2]. For general purpose 'Radiography B' is recommended. This covers spot or check radiographic examination of the welded joints in question, comprising at least 10 per cent of their whole length. The individual radiographs in general are not less than 250 mm unless shape of the joints obstructs.

Detail interpretation of radiographs and recommended practice for radiographic examination are given in IS : 1182—1967 and IS : 4853—1968. Butt welds in drums and shells and longitudinal butt welds in headers will not be acceptable, if one or more of the following defects are existing.

1. For radiography A
   (a) Cracks or areas having incomplete fusion or penetration ;
   (b) Any elongated inclusion of a length exceeding :
      (i) half the thickness with a maximum of 6 mm for thickness not exceeding 18 mm,
      (ii) one third of thickness for thicknesses over 18 mm and up to and including 75 mm, and
      (iii) 25 mm for thicknesses exceeding 75 mm.
   (c) Any group of inclusions of slag in alignment, the total length of which exceeds the thickness over a length of 12 times the thickness except when the distance between successive defects exceeds 6 times the length of the longest defect in the group ; and
   (d) Any porosity greater than that specified in codes[2].
2. For radiography B
   (a) Cracks or areas having incomplete fusion or penetration ;

(b) Any inclusion or cavities of a length exceeding 2/3 of the thickness of thinner plate welded ;

(c) Any group of inclusions in alignment, the total length of which exceeds the thickness over a length of 6 times the thickness, except when the distance between the successive defects exceeds 3 times the length of the longest defect in the group. The maximum length of elongated inclusion permitted should not be more than 12 mm ;

(d) Porosity is not a factor in the acceptability of welds not required to be fully radiographed.

Among advantages radiography provides permanent record on film, works well on thin sections, high sensitivity, fluoroscopy techniques available, adjustable energy level.

This technique has disadvantages also. Among are high initial cost, power source required, radiation hazard, trained technician needed.

13.4.3 Dye penetrant tests

Dye penetrant test procedure is elaborately given in IS : 3658 — 1966. This technique is employed in locating cracks, porosity, laps, lack of weld bond, fatigue and grinding cracks in the surface. It is simple to apply, portable, fast, low in cost, easy to interpret the results, no elaborate set up required. On the other hand this technique is limited to surface defects only ; surfaces must be clean.

This test is made by using two dyes, one red colour called 'Penetrator' and the other one white colour called 'Developer or Detector'. By this method cracks, holes, etc., can be detected not only in the weld but also in the entire surface of the vessel. The working principle is as follows :

First, red 'Penetrant Dye' is to be spreaded over the clean dry test section with the help of brush or spray. The dye is allowed to remain for 5 — 10 minutes. This permits the coloured liquid to enter into any fine opening due to capillary action. Then the red liquid is thoroughly washed out with ordinary tap water. Due to the high surface tension of the water, no coloured material from the crack or defective holes can be washed out. The surface is then dried, if necessary with hot air. To accelerate the drying process hot water can be used for washing operation. This is not necessary if sufficient time can be permitted for drying.

Now on the dry surface the thoroughly shaken white 'Developer' is applied by means of brush or spray. The white liquid then drives the entrapped red liquid out from defective spots like crack, hole, etc., and accordingly identified red marks will be observed on those places. From the nature and intensity of those red marks on the white 'Developer', one can identify the severity of defects.

The testing section should be made free from oil film. The presence of slight oil film may not disturb much the function of the 'Penetrant Dyes', but the surface must be made dry.

13.4.4 Freon tests

This method is employed to find leakages through welded joints and other places like flange joints, tube joints in tube sheet, etc. The instrument used for this test is named as "Halogen-Leakfinder".

The leak-detector with "Snuffing-device" traces the leakage in a vessel filled with halogen containing liquid, especially Freon 11 and 12.

The instrument consists of a Transformer, accessories and "Snuffer" (Portable). Indication of the leakages is obtained through loudspeaker or needle-pointer on the Leakfinder. The sensitivity of the instrument is described by the fact that it can detect a leakage through which Freon can escape at the rate of 300 mg/year.

The measuring principle utilizes the fact that the emission of positive ions from glazing platinum increases tremendously by coming in contact with halogens.

A "Snuffing-Cathode-Ray-Tube" consisting of platinum anode is placed in the portable apparatus. Since the emitting ions are Alkali ions, which are fixed on the platinum-anode as potassium atoms beforehand, after a few hundred hours of operation the alkali content of platinum becomes weak. The snuffing-cathode-ray-tube, then, loses its sensitivity very fast, and it must be replaced.

In search of leakage in an apparatus the portable unit with its tip is to be guided slowly in the leak-proof joints like weld seam, brazed joints, flange connections, stuffing box, etc.

If there is any leak, the Freon will be escaping through these points and the fan which is provided in the portable apparatus will suck the Freon through the suction tube into the snuffing-cathode-ray-tube. As a result higher amount of ions from the platinum-anode are produced in the cathode-ray-tube. The increased flow of ions is registered with the measuring instrument of the leak-finder optically or through loudspeaker.

13.4.5 Magnetic tests

IS : 3703—1966 provides code of practice for magnetic flaw detection. This method is employed for detecting surface or shallow subsurface flaws, cracks porosity, non-metallic inclusions and weld defects. This technique is employed only for ferro magnetic materials.

Flaw detection by magnetic particles is economical, simple in principle, easy to perform, portable (for field testing), fast for production testing.

Limitations are — material must be magnetic, demagnetizing after testing required, power source needed, parts must be cleaned.

13.4.6 Ultrasonic tests

IS : 3664—1966 and IS : 4260—1967 are the code of practice for ultrasonic testing by pulse echo method (direct contact) and recommended practice for ultrasonic testing of welds in ferritic steels respectively.

The technique is used for finding internal defects, cracks, lack of bond, laminations, inclusions, porosity, determining grain structure and thickness.

This method can be employed for all metals and hard non-metallic materials, sheets, tubing, rods, forgings, castings, field and production testing, in-service part testing, brazed and adhesive-bonded joints.

The ultrasonic testing procedure is fast, dependable, easy to operate, lends itself to automation, results of test immediately known, relatively portable, highly accurate, sensitive.

It requires contact or immersion of parts, interpretation of readings requires training. These are limitations.

CHAPTER 14

DESIGN OF SOME SPECIAL PARTS

14.1 INTRODUCTION

In designing certain parts of process equipment, special stress considerations are necessary. In this chapter design procedure of expansion joints for heat exchangers, expansion loop in piping system, tube plate thickness of heat transfer equipment, bubble-cap and sieve-plate thickness of distillation column will be discussed.

14.2 EXPANSION JOINT FOR HEAT EXCHANGERS[1]

Expansion joints are often found essential for fixed tube heat exchangers to accommodate differential expansion between shell and tubes, which may be expected during either start-up, shut-down, cleaning or normal operation and thus to reduce thermal stresses in the shell and tubes. The design of an expansion joint depends on both the amount of differential expansion and the cyclic conditions to be expected during operation.

Basic principle of constructing expansion joint is to provide discontinuity in the shell for its free expansion or contraction. All expansion joints are either bellows or corrugated types. Various types of expansion bellows and their service limits are discussed in Chemical Engineers Hand Book, 4th Edition.

14.2.1 Design procedure

The expansion joint is to carry axial load and load due to internal pressure, so also load due to bending moment, which a single bellow experiences due to compaction. The highest loading point is the junction between the knuckle and straight section of the bellow. This point is subject to the bending stress. Dimensions and number of bellows should be so selected that the bending stress value remains within allowable limit.

To explain the nomenclature in various equations, Fig. 14.1 is taken as reference. In any design initial dimensioning is an important criterion. Suitability of the initial dimensions for any particular system is to be checked by appropriate method of stress analysis. Dimensions for knuckle radius and length of the flat portion of a bellow are to be decided initially. To minimise the stress concentration the knuckle radius should have some minimum value depending on the thickness of the bellow plate. A correlation of the following type may be used to determine the knuckle radius, r_k.

---

[1] B. C. Bhattacharyya, "Design aspects of expansion joints for heat exchangers", Seminar proceedings on "Heat exchanger", Indian Chemical Manufacturer, 1971.

$$r_k \geqslant \frac{0.22\, D_i \sqrt{\dfrac{p\, X}{\sigma_y}}}{1 - 0.76 \sqrt{\dfrac{p\, X}{\sigma_y}}} \qquad \ldots (14.2.1)$$

Where,

p = design pressure

σ_y = yield stress of bellow material at design temp.

X = safety factor depending on elongation at rupture

= 1.3 for elongation > 25%

= 1.4 for 20 — 25%

= 1.5 for elongation < 20%

Value of r_k may be rounded of as a multiple of 5. Materials undergoing elongation less than 15% at rupture should not be used as bellows.

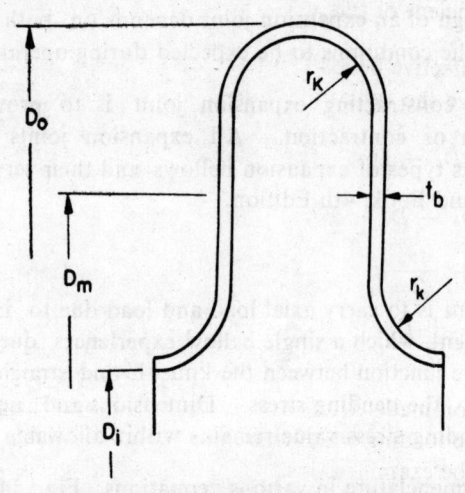

Fig. 14.1 Expansion bellow

The outer diameter, D_o, of the bellow may be estimated from the following :

$$D_o \geqslant D_i + 7\, r_k \qquad \ldots (14.2.2)$$

Where, D_i is the inner diameter of the bellow :

To estimate the bellow-wall thickness following two equations are suggested[2].

---

[2] K. Wellinger and H. Dietmann, "Festigkeitberechnung Von Wellrohrkompensatoren", Konstruktion, 17 (1965), Heft 3.

DESIGN OF SOME SPECIAL PARTS

$$t_b \geqslant 0.25 (D_o - D_i - 3 r_k) \sqrt{\frac{p \, X}{\sigma_\lambda}} \qquad \ldots (14.2.3)$$

$$t_b \geqslant \frac{p \, X}{\sigma_y} \, \frac{r_k (D_o + D_i)}{(2.3 \, r_k + D_o - D_i)} \qquad \ldots (14.2.4)$$

Larger of the two is to be taken as the wall-thickness. To that necessary corrosion allowance is to be added.

Thickness calculated using Eqs. 14.2.3 and 14.2.4 will be safe against internal pressure, p. Bellow-wall thickness, t_b is to be examined against fatigue failure.

The largest axial deformation which the expansion joint should take care of (neglecting residual stress) is given below:

$$L_A = (T_p \, \alpha_p - T_s \, \alpha_s) \, L \qquad \ldots (14.2.5)$$

Where,

L = common length for shell and tubes
T_p = temperature of pipe-tubes
T_s = temperature of shell
α_p = coefficient of thermal expansion for pipe
α_s = coefficient of thermal expansion for shell

Let L_o be the effective bellow height to flex during axial deformation. Following two equations, then, give the values for L_o and smaller of the two is to be chosen.

$$L_o = \tfrac{1}{2} \left[D_o - D_i - 2 t_b + \left(2 - 50 \, \frac{r_m}{D_m} \right) r_m \right] \qquad \ldots (14.2.6)$$

$$L_o = \tfrac{1}{2} (D_o - D_i - 2 t_b - r_m) \qquad \ldots (14.2.7)$$

Where,

$$r_m = r_k + \frac{t_b}{2}$$

$$D_m = \frac{D_o + D_i}{2}$$

Any construction made with ductile material has inherent capacity to withstand certain amount of cyclic loading without fatigue failure. This is also true in case of expansion joints. Let it be examined now, for a bellow designed by above procedure, what may be the maximum axial deformation, L_∞ which will not cause fatigue failure for any number of load cycles. An expression of the following type is suggested[2] for that.

$$L_\infty = 4 \, \frac{\sigma_w \, r_o \, L_o^2}{X_w \, E \, D_m \, t_b} \qquad \ldots (14.2.8)$$

Where,

σ_w = cyclic stress
= 0 5 (U. T. S. at design temperature) for carbon steel

X_w = safety against cyclic stress
   = 1.6
D_m = mean diameter of bellow
E = modulus of elasticity at design temperature
$r_o = \frac{1}{2}(D_m - L_o)$

If $L_A \leqslant L_\infty$, no fatigue failure is expected for the expansion joint. In practice it is not necessary to design an expansion joint for infinite number of load cycles. This will unnecessarily increase the cost for expansion joints. From fatigue stress analysis one can determine the safe number of load cycles for a L_A/L_∞ ratio.

The axial reactive load, F_A, experienced by the bellow per unit length deformation in axial direction is given by,

$$F_A = \frac{\pi}{4} E (D_o + D_i) \frac{t_b^3}{L_o^3} \qquad \ldots (14.2.9)$$

F_A can be reduced proportionately if number of bellows are increased.

The bending moment, M, caused due to axial reactive load, F_A, is estimated from,

$$M = (F_A L_A) L_o \qquad \ldots (14.2.10)$$

The approximate method of determining the bending stress across the bellow thickness at the top is to determine first the section modulus assuming rectangular cross-section of unit width and depth t_b. For this, section modulus $Z = t_b^2/6$ and bending stress, σ_b, is given by,

$$\sigma_b = \frac{M_o}{Z} = \frac{6 M_o}{t_b^2} \qquad \ldots (14.2.11)$$

Where, M_o is the bending moment per unit circumferential length (i.e., unit width).

So,
$$M_o = \frac{M}{\pi D_o} \qquad \ldots (14.2.12)$$

With this one should take account for axial thrust and hoop stress due to internal pressure and also direct compressive stress caused by axial load. For safety design maximum stress value should remain within yield-limit.

14.3 EXPANSION LOOP IN PIPING SYSTEM

It is common feature that process pipe lines are subjected to temperature changes. This causes expansion or contraction resulting considerable stresses in the piping system. In such case, if free movement of the line is restricted, the pipe wall and joints are subjected to high stresses, and large thrusts may be exerted against anchors and equipment to which the pipe line is connected. These effects are significant, though in high temperature installations, in low temperature services also these cannot be neglected. To facilitate free movement of pipe lines caused by thermal expansion or contraction, either expansion joints or changes in the direction of the line are to be adopted.

DESIGN OF SOME SPECIAL PARTS

14.3.1 Design of equations[3] for expansive forces in pipe lines

Various types of expansion joints may be designed for process piping. Expansion loop of the type shown in Fig. 14.2 will be discussed in this section.

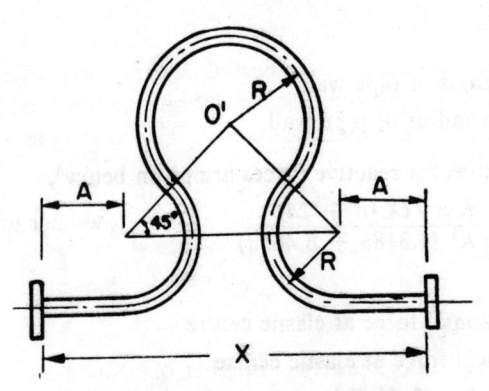

Fig. 14.2 Expansion loop (270°)

Before going to the equations to determine the induced stresses in the piping system resulting from thermal expansion, it will be logical to discuss the design philosophy of functioning of an expansion joint. Due to thermal expansion displacement occurs in the pipe line. This causes bending if the two ends are rigidly held. For a simple single-plane pipe line two forces are required to keep the line in position. If an expansion joint is present in the line, the displacement occurs in it and depending on its flexible nature, reactive forces of much less magnitude occurs in the system.

For convenience in computation, the reactive forces F_x and F_y at the expansion joint are usually referred to the elastic centre O of the system, which is the centroid of the projected areas of the pipe comprising the entire line. The location of the elastic centre O is defined by the coordinates x and y with respect to an origin O' as shown in Fig. 14.2. For this type of expansion joints, coordinates are given by the following equations.

$$x = 0 \qquad \ldots (14.3.1)$$

$$y = -R \frac{2.414\,a + 0.354\,n}{a + n/2} \qquad \ldots (14.3.2)$$

where,

a = virtual length of section of pipe

   = A (actual length for plain straight pipe)

n = virtual length of pipe bend or loop

   = $9.42\,kR$ (for 270° plain expansion loop)

---

[3] H. C. Hesse and J. H. Rushton, "Process Equipment Design", Affiliated East-West Press Pvt. Ltd., New Delhi.

R = mean radius of pipe bend
k = flexibility factor
$= \dfrac{12h^2 + 10}{12h^2 + 1}$

h = bend stiffness characteristic of plain curved pipe
$= \dfrac{tR}{r^2}$

t = thickness of pipe wall
r = mean radius of pipe wall

The equations for reactive forces are given below[3],

$$F_x = + \dfrac{E \alpha TIX(n + 2a)}{n R^2 (1.318n + 8.465a)} \qquad \ldots (14.3.3)$$

$$F_y = 0 \qquad \ldots (14.3.4)$$

where,
F_x = horizontal force at elastic centre
F_y = vertical force at elastic centre
E = modulus of elasticity of pipe material
α = coefficient of expansion
T = operating temperature
I = Plane moment of inertia of pipe cross section
$= \dfrac{\pi}{64}(d_o^4 - d_i^4)$
X = distance between pipe supports or points of fixation

14 3.2 Determination of induced stresses in the pipe

The stress σ_f in the pipe due to bending in the plane of the system is given by,

$$\sigma_f = \dfrac{iM}{Z} \qquad \ldots (14.3.5)$$

where,
i = stress intensification or concentration factor
$= \sqrt{\dfrac{(12h^2 + 10)^3}{9(12h^2 + 1)}}$ (for $h \leqslant 1.472$)
$= \dfrac{12h^2 - 2}{12h^2 + 1}$ (for $h > 1.472$)
$= 1.0$ (for plain straight pipe)
M = bending moment
Z = section modulus of pipe cross section
$= \dfrac{I}{r_o}$

DESIGN OF SOME SPECIAL PARTS   229

The moment at any point in a system is equal to the algebraic sum of the products of the forces F_x and F_y and the distances of the elastic centre from the point under the consideration.

The longitudinal stress in the pipe wall induced by the internal pressure is given by,

$$\sigma_p = \frac{p\,r}{2\,t} \qquad \ldots (14.3.6)$$

The total maximum longitudinal stress σ_m is given by,

$$\sigma_m = \sigma_f + \sigma_p \qquad \ldots (14.3.7)$$

The allowable unit stress f, which would be equal to or greater than σ_m, is given by

$$f = 0.75\,(f_a + f_o) \qquad \ldots (14.3.8)$$

where,

f_a = permissible stress at atmospheric temperature

f_o = permissible stress at operating temperature

Design Example 14.1 : A steam line from boiler house to the unit operation laboratory is to be installed. From the following data decide if the wall thickness and the expansion loop radius are properly selected.

Data :

O.D. of pipe	...	60 mm
I.D. of pipe	...	50 mm
Material of pipe	...	IS : 1979 — 1961 St 30
Steam pressure	...	1.0 MN/m²
Steam temperature	...	185 °C
Corrosion allowance	...	2 mm
Mean radius of expansion loop	...	300 mm
$E \propto T$ for pipe material	...	380 MN/m²
Distance between two anchor points	...	50 m

Solution : Here the solution is to determine the maximum induced stress in the system and to compare with the maximum allowable stress value of the pipe material. If the induced stress is found less than the allowable stress, the piping system will be safe, if the pressure-temperature fluctuations are not too frequent.

For this construction,

$2A = X - 3R$

$\quad = 50 - 3\,(0.3)$

$\quad = 49.1$ m

So, $A = 24.55$ m $= a$

To determine y from Eq. 14.3.2 and F_x from Eq. 14.3.3 various parameters are evaluated at first.

$$h = \frac{t\,R}{r^2}$$

$$= \frac{5 \times 300}{(27.5)^2} = 2.0$$

$$k = \frac{12\,h^2 + 10}{12\,h^2 + 1}$$

$$= \frac{12\,(2.0)^2 + 10}{12\,(2.0)^2 + 1} = 1.18$$

$$n = 9.42\,k\,R$$
$$= 9.42 \times 1.18 \times 0.3 \text{ m}$$
$$= 3.35 \text{ m}$$

$$I = \frac{\pi}{64}(d_o^4 - d_i^4)$$

$$= \frac{\pi}{64}\left[(0.06)^4 - (0.05)^4\right] \text{m}^4$$

$$= 3.3 \times 10^{-7} \text{ m}^4$$

Substituting,

$$y = -0.3 \times \frac{2.414\,(24.55) + 0.354\,(3.35)}{24.55 + 3.35/2}$$

$$= -0.7 \text{ m}$$

$$F_x = 380 \times 3.3 \times 10^{-7} \times \frac{50\,(3.35 + 2 \times 24.55)}{3.35\,(0.3)^2\,(1.318 \times 3.35 + 8.465 \times 24.55)}$$

$$= 5.15 \times 10^{-3} \text{ MN}$$
$$= 5\,150 \text{ N}$$

Bending moment due to load F_x will be maximum at the top section of the loop. The moment arm Y from the elastic centre is given by.

$$Y = 0.7 + 0.3$$
$$= 1.0 \text{ m}$$

So, $M = F_x \times Y$
$$= 5\,150 \times 1.0$$
$$= 5\,150 \text{ N m}$$

$$Z = \frac{I}{r_o}$$

$$= \frac{3.3 \times 10^{-7}}{0.03} \text{ m}^3$$

$$= 1.1 \times 10^{-5} \text{ m}^3$$

DESIGN OF SOME SPECIAL PARTS

For $h = 2.0$

$$i = \frac{12(2.0)^2 - 2}{12(2.0)^2 + 1}$$

$$= 0.94$$

Substituting in Eq. 14.3.5,

$$\sigma_f = \frac{0.94 \times 5\,150}{1.1 \times 10^{-5}} \text{ N/m}^2$$

$$= 4.4 \times 10^8 \text{ N/m}^2$$

$$= 440 \text{ MN/m}^2$$

From Eq. 14.3.6

$$\sigma_p = \frac{p\,r}{2\,t}$$

$$= \frac{1.0 \times 27.5}{2 \times 5}$$

$$= 2.75 \text{ MN/m}^2$$

From Eq. 14.3.7

$$\sigma_m = \sigma_f + \sigma_p$$

$$= 440 + 2.75$$

$$= 442.75 \text{ MN/m}^2$$

From Eq. 14.3.8

$$f = 0.75\,(f_a + f_o)$$

Where, $f_a = f_o = 135 \text{ MN/m}^2$

Therefore, $f = 0.75\,(2 \times 135)$

$$= 202.5 \text{ MN/m}^2$$

As the allowable unit stress f is less than the maximum longitudinal stress σ_m, the selection of thickness and loop size is not appropriate. It may be noted that thermal stress is very large compared to the pressure stress. The system should also be checked under corroded condition, when flexibility increases due to reduction of thickness, though pressure stress increases.

14.4 Tube plate thickness

In this section three cases of tube plate thickness calculation will be considered. These are : (a) Fixed tube plate ; (b) Tube plate with hair-pin U-tube and (c) Tube plate with floating head.

Stresses in the tube plates are caused due to pressure difference in the shell and tube sides and also due to differential thermal expansion. These cause radial as well as bending tensile or compressive stresses. Tube plate thickness should be so decided that neither of these induced stresses exceeds the permissible stress value of the plate material at the design temperature. Also to be noted that for economic reason tube plate thicknesses should not be too large. This condition is achieved by adjusting the thickness in such a way that at least one of the induced stress values is close to the allowable stress value. In this connection attention is drawn to Table 14.1, where minimum plate thickness is specified[4]. If the calculated thickness comes less than the specified minimum value, the thickness given in the table is to be taken. Before comparing with the Table 14.1, corrosion allowance and depth of groove for partition plate, if any, are to be added with the calculated plate thickness. In calculating plates thickness, two kinds of attachments are considered. These are important for determining induced stresses.

(a) Clamped — In this type of attachments plates are either welded with the shell or bolted with full face, so that angular deflection at the periphery is effectively prevented.

(b) Simply supported — Here the tube plate is held within the bolt holes of the flange joint.

The effects of these two conditions are presented in Table 14.2 in three dimensionless functions G_1, G_2 and G_3 as a dependent variable of a dimensionless factor k given in Eq. 14.4.1. G_1, G_2 and G_3 are required for determining the induced stress values.

Table 14.1 Minimum Tube Plate Thicknesses[4]

Tube outer dia (mm)	Thickness (mm)
6	6
10	10
12	12
16	13
18, 19, 20	15
25, 25.4	19
31.8, 32	22.4
38, 40	25.4

[4] IS : 4503 — 1967 : "Specification for shell and tube type heat exchangers", Indian Standards Institution, New Delhi.

DESIGN OF SOME SPECIAL PARTS

Table 14.2 Values[1] for G_1, G_2 and G_3 as Function of k

k	G_1 clamped	G_1 simply supported	G_2 clamped	G_2 simply supported	G_3 clamped	G_3 simply supported
0	1.28	0.78	+ 1.00	+ 1.00	1.00	1.00
1	1.30	0.80	+ 0.99	+ 0.97	1.00	1.05
2	1.39	1.05	+ 0.84	+ 0.50	1.10	1.40
3	1.55	1.88	+ 0.37	− 0.36	1.35	2.25
4	1.90	2.70	− 0.15	− 0.74	1.70	3.10
5	2.38	3.50	− 0.33	− 0.62	2.06	3.80
6	2.90	4.20	− 0.27	− 0.52	2.40	4.55
7	3.40	4.98	− 0.22	− 0.53	2.75	5.30
8	3.80	5.65	− 0.20	− 0.55	3.11	6.00
9	4.32	6.38	− 0.21	− 0.60	3.45	6.70
10	4.80	7.10	− 0.22	− 0.64	3.82	7.40
11	5.25	7.80	− 0.23	− 0.69	4.20	8.15
12	5.71	8.54	− 0.25	− 0.73	4.50	8.85
13	6.20	9.20	− 0.27	− 0.78	4.82	9.52
14	6.65	9.85	− 0.28	− 0.82	5.20	10.22
15	7.15	10.63	− 0.29	− 0.87	5.56	10.95
16	7.60	11.40	− 0.31	− 0.91	5.90	11.65
17	8.08	12.06	− 0.32	− 0.95	6.32	12.40
18	8.57	12.80	− 0.33	− 1.00	6.65	13.05
19	9.00	13.50	− 0.35	− 1.04	6.98	13.80
20	9.48	14.20	− 0.36	− 1.08	7.25	14.40

The value of k is to be obtained from the following equation.

$$k^2 = 1.32 \frac{D}{t_p} \sqrt{\frac{n\,a\,A}{(A-C)\,L\,t_p}} \qquad \ldots (14.4.1)$$

Where,

A = cross-sectional area of bore of shell

$\quad = \dfrac{\pi D^2}{4}$

a = cross-sectional area of metal in one tube

C = cross-sectional area of tube holes in tube plate

$$= \frac{n \pi d^2}{4}$$

D = bore of cylindrical shell
d = outside diameter of tube
L = effective length of the tubes
n = number of tubes
t_p = thickness of the tube plate, excluding corrosion allowance.

14.4.1 Fixed tube plate

The maximum radial stress σ_r induced in tube plate and maximum longitudinal stress σ_t induced in tubes are expressed by the following correlations[4]. Value of σ_r may either be positive or negative depending upon the side of the plate to which the stress applies. For σ_t positive value denotes tensiles stress and negative value compressive stress.

$$\sigma_r = \frac{\left(\frac{A-C}{A}\right)\left(\frac{D}{t_p}\right)^2 p_b}{4 \mu G_1 \left(\frac{E_t\, n\, a}{E_s\, B} + G_3\right)} \quad \ldots (14.4.2)$$

$$\sigma_t = \frac{A-C}{n\,a}\left[p - \frac{p_b\, G_2}{\frac{E_t\, n\, a}{E_s\, B} + G_3}\right] \quad \ldots (14.4.3)$$

$$\text{or,} \quad = \frac{A-C}{n\,a}\left[p - \frac{p_b\, G_3}{\frac{E_t\, n\, a}{E_s\, B} + G_3}\right] \quad \ldots (14.4.3a)$$

whichever is larger.

In these expressions:

B = cross-sectional area of shell plate
$\quad = \pi(D+t)t$
E_s = modulus of elasticity for shell
E_t = modulus of elasticity for tubes
p = equivalent pressure difference
$\quad = p_1 - p_2 - \frac{n\,a}{A-C} p_2$
p_1 = gauge pressure outside tubes
p_2 = gauge pressure inside tubes

$$p_b = p_1 - p_2\left[1 + \frac{n\,a}{A-C} + \frac{E_t\, n\, a\, A}{E_s\, B(A-C)}\right] + \frac{n\, a\, E_t\, y}{A-C}$$

t = thickness of shell
y = total differential strain
   = $\alpha_t (\theta_t - \theta_r) - \alpha_s (\theta_s - \theta_r)$
α_s = coefficient of linear expansion of the shell per deg C
α_t = coefficient of linear expansion of the tube per deg C
μ = ligament efficiency assumed equal to 0.4 for single-pass heat exchangers, 0.5 for two-pass heat exchangers, 0.6 for four-pass heat exchangers in the absence of any better alternative.
θ_s = temperature of the shell
θ_t = temperature of the tubes
θ_r = assembly temperature

It is to be noted that simultaneous values of p_1, p_2 and y which may occur in service giving rise to the highest numerical value of p_b should be taken for design calculations.

14.4.2 Tube plate with floating head

Following equations are suggested[4] to determine σ_r and σ_t.

$$\sigma_r = \frac{p_1 - p_2}{4 \mu G_1} \left(\frac{D}{t_p}\right)^2 \qquad \ldots (14.4.4)$$

$$\sigma_t = \frac{A - C}{n a} \left[p - \frac{(p_1 - p_2) A G_2}{A - C} \right] \qquad \ldots (14.4.5)$$

or, $= \dfrac{A - C}{n a} \left[p - \dfrac{(p_1 - p_2) A G_3}{A - C} \right] \qquad \ldots (14.4.5a)$

Whichever is larger.

Simultaneous values of p_1 and p_2 which may occur in service giving rise to the highest numerical value of $p_1 - p_2$ should be considered for design calculations.

14.4.3 Tube plate with hair-pin U-tube

Expressions for radial stress σ_r and longitudinal stress σ_t are given by the following equations[4].

In case of simply supported

$$\sigma_r = \frac{0.309 (p_1 - p_2)}{\mu} \left(\frac{D}{t_p}\right)^2 \qquad \ldots (14.4.6)$$

In case of clamped tube plate

$$\sigma_r = \frac{0.187 5 (p_1 - p_2)}{\mu} \left(\frac{D}{t_p}\right)^2 \qquad \ldots (14.4.7)$$

For both the cases

$$\sigma_t = \frac{- (p_1 - p_2) C}{n a} - p_2 \qquad \ldots (14.4.8)$$

Solutions of practically all the equations require trial and error method. Plate thickness t_p is to be assumed and on that basis σ_r and σ_t are to be determined. Neither of these two values, though, must not exceed the allowable stress values of the materials, at least one value should be as close as possible to the allowable stress value.

14.5 BUBBLE CAP TRAY DESIGN

The mechanical design of a bubble cap tray requires following considerations:

1. The dead weight of the tray with its accessories.
2. The dynamic loading due to varying liquid and vapour loads imposing flexure stresses on the tray.
3. Accommodation of thermal expansion if the operating temperature is significantly different from atmospheric.
4. The load imposed by maintenance men on the trays, if any.

Besides the above considerations following points are also to be taken into account:

1. The means of cleaning individual trays, if required.
2. The means for periodic inspection of the trays and internal tower wall.
3. The provision of access to enable individual tray designs to be altered in a minor fashion due to change in design conditions.

14.5.1 Factors affecting tray design

It is essential that the tray floor remains flat under all design conditions. Possible deflection from liquid loading, thermal expansion or vapour thrust, plus the weight of the tray itself must be taken care of. For the larger diameter sectional trays the supporting framework is designed separately from the tray floor, as is designed to take the full loads under dynamic conditions with a maximum deflection of 1 mm per meter span[5]. The tray is then of relatively thin material supported on the framework of support beams. The tray itself must, however, be designed to take a dynamic load over its unsupported area with a certain minimum deflection (say, 1 in 400).

The sections are simply supported at the corners, thermal expansion being allowed for by attaching the sections to the support beams by friction clamps.

It is generally conceded that the reinforcement provided by either risers welded in or pressed in does allow the tray section to be designed as a flat plate simply supported at the corners and subject to a dynamic load.

An allowable stress for the material should be used corresponding to that recommended for the working temperature of the tray material.

---

[5] F. Molyneux, "Chemical Plant Design — I", Butterworths, London, 1963.

14.5.2 Design equations

Following equations are based on the small deflection theory of flat plates[6]. Deflection is considered small if the maximum deflection of the plate under load does not exceed about one-half the thickness. From the point of performance characteristics, deflection in bubble plate must be as low as possible. Again, the magnitude of induced radial and tangential stresses, as well as the amount of deflection will depend upon the manner of loading and edge conditions. Manner of loading, however, may be considered as uniform load over entire surface of the plate. Ideally two edge conditions are simply supported and fixed edges. The condition of simply supported edges is achieved if the edge rotation of the plate is not restricted by any means. In case of fixed edges such rotation is prevented. Theoretical equations are derived for either of these two ideal conditions of support. In actual construction simply supported edges of bubble plate are not practical. In that case the bubble plates are to be placed simply on support rings without any joints. This is not generally done. The condition of fixed edges is realized by either welding the plate with the shell or by bolting between two flanges with full face. On the other hand, if the plate is supported between two flanges with narrow face, i.e. within bolt holes, or on support rings with bolted joints, the condition cannot be considered as truly fixed edges. In such cases a certain amount of edge yielding or relaxation of fixing moments occurs. For this reason it is usually advisable to design a fixed-edged plate that is to carry uniform load for somewhat higher centre stresses than are indicated by the equations. This can usually be taken into account by means of the correlations for edge slope and the principle of super-position, which will be explained with a design example at the end of this section.

It is to be noted that the following equations are valid for uniform load over entire surface of a circular plate under specified edge conditions.

Case A : Edges simply supported

In this case the deflection y will be maximum at the centre of the plate and the expression is given below.

$$y\ (max.) = \frac{3\ W\ (m-1)\ (5m+1)\ R^2}{16\ \pi\ E\ m^2\ t_p^3} \qquad \ldots (14.5.1)$$

Where, W = total applied load
  = $w\ \pi\ R^2$
  w = unit applied load
  R = radius of the plate upto support edge
  E = modulus of elasticity of plate material
  t_p = thickness of plate
  m = reciprocal of Poisson's ratio (μ)

---

[6] R. J. Roark, "Formulas for Stress and Strain", McGraw-Hill Book Company, New York.

Unit stresses are also maximum at the centre of the plate as given below.

$$\sigma_r \text{ (max.)} = \sigma_t \text{ (max.)} = - \frac{3W}{8 \pi m t_p^2} (3m + 1) \qquad \ldots (14.5.2)$$

Where,

σ_r = unit stress in the radial direction

σ_t = unit stress in the tangential direction

Negative sign for stresses indicates compression at upper surface and equal tension at lower surface. Positive sign indicates reverse condition.

Slope θ (rad) at the edge of plate measured from horizontal is given by,

$$\theta = \frac{3W(m-1)R}{2 \pi E m t_p^3} \qquad \ldots (14.5.3)$$

Case B. Edges fixed

Under such condition the maximum deflection at the centre is given by,

$$y = \frac{3W(m^2 - 1)R^2}{16 \pi E m^2 t_p^3} \qquad \ldots (14.5.4)$$

In this case the unit stress in the radial direction is maximum at the edge and is given by,

$$\sigma_r \text{ (max)} = \frac{3W}{4 \pi t_p^2} \qquad \ldots (14.5.5)$$

The expression for unit stress in the tangential direction induced at the edge is as follows:

$$\sigma_t = \frac{3W}{4 \pi m t_p^2} \qquad \ldots (14.5.6)$$

Both the unit stresses at the centre are equal and given below.

$$\sigma_r = \sigma_t = - \frac{3W(m+1)}{8 \pi m t_p^2} \qquad \ldots (14.5.7)$$

Case C. Edges not truely fixed

This is the case of superposition as discussed earlier in this section. Here the edges cannot be considered as ideally fixed and therefore, a certain amount of edge yielding occurs resulting in a small slope at the edge and relaxation of moments at the edge.

Under such condition the maximum deflection at the centre is to be determined from the following relationship.

$$y = \frac{3W(m^2 - 1)R^2}{16 \pi E m^2 t_p^3} + \frac{6(m-1)MR^2}{E m t_p^3} \qquad \ldots (14.5.8)$$

In this expression M is the amount of relaxation of the fixing moments at the edge; or on the other hand it can be said that M is the uniform edge moment necessary to cause an edge slope θ' as shown below.

DESIGN OF SOME SPECIAL PARTS

$$\theta' = \frac{12(m-1)MR}{E\,m\,t_p^3} \qquad \ldots (14.5.9)$$

Uniform edge moment M is to be determined from Eq. 14.5.9 by assuming a small value for θ'. The magnitude of edge slope θ' will depend upon the nature of the edge-fixing. However, the maximum value will not exceed θ given in Eq. 14.5.3. Once M is obtained, the unit stresses at different places can be estimated by using following equations.

At the edge

$$\sigma_r = \frac{3W}{4\pi t_p^2} - \frac{6M}{t_p^2} \qquad \ldots (14.5.10)$$

$$\sigma_t = \frac{3W}{4\pi m t_p^2} - \frac{6M}{t_p^2} \qquad \ldots (14.5.11)$$

At the centre

$$\sigma_r = \sigma_t = -\left[\frac{3W(m+1)}{8\pi m t_p^2} + \frac{6M}{t_p^2}\right] \qquad \ldots (14.5.12)$$

It may be mentioned that even a small horizontal force at the line of contact may appreciably reduce the stress and deflection in a supported plate, while on the other hand a very slight yielding at nominally fixed edges will greatly relieve the stresses there, while increasing the deflection and centre stresses. These are seen from Eqs. 14.5.8, 14.5.10 to 14.5 12.

Case D: Effect of temperature variation

All the expressions for stresses presented in earlier paragraphs do not take into account the thermal stresses. In practice, the fabrication temperature and the operating temperature may not be so close that the temperature effect in calculating stresses can be neglected. However, the magnitude of thermal stresses will depend upon the manner of supporting the plate. In case of simply supported plate no stresses are set up since its thermal expansion may not be restricted externally or internally. But if the plates are bolted or welded preventing their free expansion, thermal stresses would develop; or expansion can only be taken care of by flexing the plate and disturbing the design characteristics. The unit thermal stress in compression can be evaluated from the following stress-strain relationship.

$$\sigma_{temp} = (\alpha_p - \alpha_s)\,\Delta T\,E \qquad \ldots (14.5.13)$$

Where,

σ_{temp} = unit stress due to temperature variation

ΔT = difference between fabrication or ambient temperature and maximum operating temperature

E = modulus of elasticity of plate material

α_p = coefficient of thermal expansion for plate

α_s = coefficient of thermal expansion for shell or support

It is to be noted that the thermal stresses are self-limiting. It gets satisfied with some deformation occurring in the body. Eq. 14.5.13 shows that thermal stress does not depend on thickness of the plate.

14.5.3 Design of sectional trays

For larger sizes the most versatile and the most frequently used is the sectional tray. This type can readily be dismantled into sections capable of passing through a manway or manhole in the wall of the vessel. Since the sections are built light enough to enable manual handling, it is necessary to use separate load-bearing members like beam and truss supports. The floor plates are laid on top of the support beams and the peripheral tray ring, using asbestos jointing between each of the members. The sections may be either square or rectangular.

A. Square sections

If the support beams are so arranged that each unsupported section becomes a square of length "a", then the following equation can be used to determine the stress and deflection of unsupported portion[6]. Here assumptions are, — edges supported below only and uniform load over entire surface.

At centre, on diagonal section :

$$\sigma = - \frac{0.2214 \, w \, a^2}{t_p^2} \qquad \ldots (14.5.14)$$

$$y \, (max.) = - \frac{0.0443 \, w \, a^4}{E \, t_p^3} \qquad \ldots (14.5.15)$$

At corners, on diagonal section :

$$\sigma \, (max.) = - \frac{0.2778 \, w \, a^2}{t_p^2} \qquad \ldots (14.5.16)$$

B. Rectangular sections

In case of rectangular sections the stress and deflection are calculated from the following correlations[6]. Assumptions are all edges simply supported and uniform load over entire surface of the section. Larger length is "a" and smaller length "b".

Maximum stress and deflection will be at the centre. The expressions are :

$$\sigma = \beta \, \frac{w \, b^2}{t_p^2} \qquad \ldots (14.5.17)$$

$$y = \alpha \, \frac{w \, b^4}{E \, t_p^3} \qquad \ldots (14.5.18)$$

The values of α and β are to be taken from the following table :

$\frac{a}{b}$	1.0	1.2	1.4	1.6	1.8	2.0	3.0	4.0	5.0	∞
β	0.2874	0.3762	0.4530	0.5172	0.5688	0.6102	0.7134	0.7410	0.7476	0.750
α	0.0444	0.0616	0.0770	0.0906	0.1017	0.1110	0.1335	0.1400	0.1417	0.1421

14.5.4 Design for larger deflection

If the maximum deflection of the tray exceeds half the plate thickness, the earlier equations will not give correct solutions. In that case following equations are to be used for stresses and deflection[6].

A. Circular plate, uniform load, edges simply supported (neither fixed nor held)

$$\frac{w R^4}{E t_p^4} = \frac{64}{63(1-\mu)}\left(\frac{y}{t_p}\right) + 0.376\left(\frac{y}{t_p}\right)^3 \qquad \ldots (14.5.19)$$

At centre:

$$\sigma = \frac{E t_p^2}{R^2}\left[\frac{1.238}{1-\mu}\left(\frac{y}{t_p}\right) + 0.294\left(\frac{y}{t_p}\right)^2\right] \qquad \ldots (14.5.20)$$

B. Circular plate, uniform load, edges fixed but not held (no edge tension)

$$\frac{w R^4}{E t_p^4} = \frac{16}{3(1-\mu^2)}\left(\frac{y}{t_p}\right) + \frac{6}{7}\left(\frac{y}{t_p}\right)^3 \qquad \ldots (14.5.21)$$

At edge:

$$\sigma = \frac{E t_p^2}{R^2}\left[\frac{4}{1-\mu^2}\left(\frac{y}{t_p}\right)\right] \qquad \ldots (14.5.22)$$

At centre:

$$\sigma = \frac{E t_p^2}{R^2}\left[\frac{2}{1-\mu}\left(\frac{y}{t_p}\right) + \frac{1}{2}\left(\frac{y}{t_p}\right)^2\right] \qquad \ldots (14.5.23)$$

C. Circular plate, uniform load, edges fixed and held

$$\frac{w R^4}{E t_p^4} = \frac{16}{3(1-\mu^2)}\left[\left(\frac{y}{t_p}\right) + 0.488\left(\frac{y}{t_p}\right)^3\right] \qquad \ldots (14.5.24)$$

At edge:

$$\sigma = 4.40\, E\left(\frac{y\, t_p}{R^2}\right) + 0.476\, E\left(\frac{y}{R}\right)^2 \qquad \ldots (14.5.25)$$

At centre:

$$\sigma = 2.86\, E\left(\frac{y\, t_p}{R^2}\right) + 0.976\, E\left(\frac{y}{R}\right)^2 \qquad \ldots (14.5.26)$$

D. Square plates under uniform load ; edges held, not fixed ; maximum stress at the centre of plate.

$\dfrac{w a^4}{E t_p^4}$	0	12.5	25	50	75	100	125	150	175	200	250
$\dfrac{y}{t_p}$	0	0.43	0.65	0.93	1.13	1.26	1.37	1.47	1.56	1.63	1.77
$\dfrac{\sigma a^2}{E t_p^2}$	0	3.80	5.80	8.70	10.90	12.80	14.3	15.6	17.0	18.2	20.5

E. Rectangular plates under uniform load; edges held, not fixed; maximum stress at the centre of plate; a/b ranges from 1.5 to ∞.

$\dfrac{w\,b^4}{E\,t_p^4}$	0	12.5	25	50	75	100	125	150	175	200	250
$\dfrac{y}{t_p}$	0	0.625	0.879	1.18	1.37	1.53	1.68	1.77	1.88	1.96	2.12
$\dfrac{\sigma\,b^2}{E\,t_p^2}$	0	4.87	7.16	10.3	12.5	14.6	16.4	18.0	19.4	20.9	23.6

Design Example 14.2: For a fractionating tower of 2 m diameter, bubble cap trays are to be designed. Weight of liquid on each tray during operation will not exceed 3 000 N. If the edges are fixed but not rigidly held horizontal under loading condition, so that an edge slope of 0.25 degree occurs, determine the thickness of the tray for maximum permissible deflection of 5 mm. Material of construction for the trays is alloy steel with allowable stress 140 MN/m² and modulus of elasticity 190 000 MN/m².

Solution: This is the Case C of section 14.5.2. Solution will be found by trial and error.

From Eq. 14.5.9, M is found.

$$\theta' = \frac{12\,(m-1)\,MR}{E\,m\,t_p^3}$$

Where, $\theta' = 0.25$ deg. $= 0.004\,36$ radian

$\dfrac{m-1}{m} = 0.7$ (for steel $\mu = 0.3$)

$R = 1.0$ m

$E = 190\,000$ MN/m²

So, $M = \dfrac{(0.004\,36)\,(190\,000)\,(t_p^3)}{(12)\,(0.7)\,(1.0)}$

$= 98.63\,t_p^3$ MN m

From Eq. 14.5.8,

$$y = \frac{3\,W\,(m^2-1)\,R^2}{16\,\pi\,E\,m^2\,t_p^3} + \frac{6\,(m-1)\,M\,R^2}{E\,m\,t_p^3}$$

Where, W = (weight of the bubble cap tray) + (weight of liquid on it)

$\dfrac{m^2-1}{m^2} = 0.91$

Weight of the bubble cap tray consists of weight of the perforated plate plus weight of the risers, caps, down comers, weir, etc. As the detail dimensions for all the accessories are not given, it may be assumed that the weight of the unpierced plate represents the tray weight. Therefore,

DESIGN OF SOME SPECIAL PARTS

$$W = \pi R^2 t_p \gamma + 3\,000$$
$$= 3.14\,(1.0)^2\,(t_p)\,(77\,000) + 3\,000$$
$$= 243\,000\,t_p + 3\,000,\ N$$

Substituting the values in consistent units

$$5 \times 10^{-3} = \frac{(3)\,(243\,000\,t_p + 3\,000)\,(0.91)\,(1)^2}{(16)\,(3.14)\,(1.9 \times 10^{11})\,(t_p{}^3)}$$
$$+ \frac{(6)\,(0.7)\,(98.63\,t_p{}^3)\,(1)^2}{(190\,000)\,(t_p{}^3)}$$

or, $3.286 \times 10^6\,t_p{}^3 = 1 + 81\,t_p$

Solving this, plate thickness t_p is found to be 8×10^{-3} m or 8 mm. It may be noted here that the maximum deflection (5 mm) is a little over half the plate thickness. Although the difference is not much, it will be better to check the thickness using equation from section 14.5.4 for large deflection.

From Eq. 14.5.21

$$\frac{w\,R^4}{E\,t_p{}^4} = \frac{16}{3\,(1-\mu^2)}\left(\frac{y}{t_p}\right) + \frac{6}{7}\left(\frac{y}{t_p}\right)^3$$

Here, $w =$ load per unit area
$$= \frac{W}{\pi R^2}$$
$$= 957\,(1 + 81\,t_p),\ N/m^2$$
$R = 1.0$ m
$E = 1.9 \times 10^{11}\ N/m^2$
$y = 5 \times 10^{-3}$ m
$\mu = 0.3$

Substituting and simplifying

$$5.86 \times 10^6\,t_p{}^3 - 59.6\,t_p = 1$$

and this gives $t_p = 6.2 \times 10^{-3}$ m or 6.2 mm

According to large deflection theory a plate thickness of 6.2 mm without corrosion allowance will satisfy the deflection requirement. Now stresses are to be checked for this thickness.

From Eq. 14.5.22 edge stress is given by,

$$\sigma = \frac{E\,t_p{}^2}{R^2}\left[\frac{4}{1-\mu^2}\left(\frac{y}{t_p}\right)\right]$$
$$= \frac{1.9 \times 10^5\,(6.2 \times 10^{-3})^2}{(1)^2}\left[\frac{4}{1-0.09}\left(\frac{5 \times 10^{-3}}{6.2 \times 10^{-3}}\right)\right]$$
$$= 25.3\ MN/m^2$$

from Eq. 14.5.23 stress at the centre is given by,

$$\sigma = \frac{E t_p^2}{R^2}\left[\frac{2}{1-\mu}\left(\frac{y}{t_p}\right) + \frac{1}{2}\left(\frac{y}{t_p}\right)^2\right]$$

$$= \frac{1.9 \times 10^5 (6.2 \times 10^{-3})^2}{(1)^2}\left[\frac{2}{(1-0.3)}\left(\frac{5 \times 10^{-3}}{6.2 \times 10^{-3}}\right) + \frac{1}{2}\left(\frac{5 \times 10^{-3}}{6.2 \times 10^{-3}}\right)^2\right]$$

$$= 20.4 \text{ MN/m}^2$$

It is quite interesting to observe that in tray design deflection is the controlling parameter.

Design Example 14.3 : If the tray of Example 14.2 is supported by beam structure, so that the maximum unsupported area does not exceed 0.5 m square, determine the required plate thickness for maximum deflection of 3 mm.

Solution : Equations given for sectional tray design are to be used in this case.

From Eq. 14.5.15,

$$y\,(max) = \frac{0.044\,3\,w\,a^4}{E\,t_p^3}$$

Where, $y = 3 \times 10^{-3}$ m

$w = 957\,(1 + 81\,t_p)$, N/m^2

$a = 0.5$ m

$E = 1.9 \times 10^{11}$ N/m^2

Substituting and simplifying,

$$2.151 \times 10^8\,t_p^3 - 81\,t_p = 1$$

By solving, plate thickness t_p is found to be 1.8×10^{-3} m or 1.8 mm. Here it is observed that the deflection is very much larger than half the plate thickness, and therefore, Eq. 14.5.15 is not appropriate in this case. Informations given in paragraph D of section 14.5.4 are to be used for the solution. Values of 3 dimensionless parameters are given in tabular form interrelating each other. The procedure will be as follows :

Assume a value for t_p and corresponding to this determine the magnitude of the first group $w\,a^4/E\,t_p^4$. Find y/t_p and check for y. Similarly estimate the value for σ.

In the given problem let $t_p = 1.8 \times 10^{-3}$ m.

Then, $\dfrac{w\,a^4}{E\,t_p^4} = \dfrac{957\,(1 + 81\,t_p)\,(0.5)^4}{1.9 \times 10^{11}\,t_p^4}$

$= 36$

Corresponding to this,
$y/t_p = 0.77$

So, $y = 0.77 \times 1.8 \times 10^{-3}$
$= 1.4 \times 10^{-3}$ m or 1.4 mm

It is interesting to note that according to large deflection theory deflection is only 1.4 mm for a plate thickness of 1.8 mm. Correspondingly $\sigma a^2/E t_p^2 = 7.0$, from which stress σ is estimated to be 17.2 MN/m².

14.5.5 Design of peripheral tray support ring

Very often trays are supported on peripheral ring. Outer edge is generally welded or riveted with the shell and inner edge free. In such case complete tray load is carried by the ring support. This condition is comparable with a circular plate with concentric circular hole having outer edge fixed and inner edge free with uniform load on concentric circular ring of radius r_m. For this, correlations are available for stresses and maximum deflection[6].

Let R = outer radius of ring support
r = inner radius of ring support
r_m = mean radius

Stress at inner edge is given by,

$$\sigma_t = \frac{-3W}{4\pi m t_p^2}\left[(m+1)\left(2\ln\frac{R}{r_m} + \frac{r_m^2}{R^2} - 1\right)\right]$$
$$- \frac{6M}{t_p^2}\left[\frac{R^2(m-1) - r^2(m+1)}{R^2(m-1) + r^2(m+1)}\right] \quad \ldots (14.5.27)$$

Stress at outer edge is given by,

$$\sigma_r = \frac{3W}{2\pi t_p^2}\left(1 - \frac{r_m^2}{R^2}\right) + \frac{6mM}{t_p^2}\left[\frac{2r^2}{R^2(m-1) + r^2(m+1)}\right]$$
$$\ldots (14.5.28)$$

$$y(max) = \frac{3W(m^2-1)}{2\pi E m^2 t_p^2}\left[\frac{(R^2 + r_m^2)(R^2 - r^2)}{2R^2} - (r^2 + r_m^2)\ln\frac{R}{r_m}\right.$$
$$\left. + (r^2 - r_m^2)\right] + \frac{6M(m^2-1)}{E m t_p^3}\left[\frac{r^4 + 2R^2 r^2 \ln\frac{R}{r} - R^2 r^2}{r^2(m+1) R^2(m-1)}\right]$$
$$\ldots (14.5.29)$$

Where, $M = \dfrac{W}{8\pi m}\left[(m+1)\left(2\ln\dfrac{R}{r_m} + \dfrac{r_m^2}{R^2} - 1\right)\right]$ ... (14.5.30)

Ring plate thickness should be such that the maximum deflection must not exceed half the plate thickness. It is also necessary to check the plate thickness against direct shearing. The induced shear stress may be expressed by the following relationship.

$$\tau = \frac{W}{2\pi R t_p} \qquad \ldots (14.5.31)$$

Allowable shear stress may be taken 80% of permissible tensile stress.

14.5.6 Design of beam support

It is mentioned earlier that larger diameter trays are supported on peripheral ring in combination with beam structures to reduce the deflection in trays. Beams of various cross-sections are used for this purpose. Selection of particular section and size will depend upon the magnitude of the load it is to carry and also on its flexural properties. As usual deflection characteristic is the controlling factors. Beams are considered to be fixed at both the ends and load is uniformly distributed. Deflection and moment equations are available in the literature[6].

$$y\,(max) = \frac{1}{384}\frac{W l^3}{E I} \qquad \ldots (14.5.32)$$

Where, W = load acting on a particular beam

l = length of the beam

E = modulus of elasticity of beam material

I = moment of inertia of the section of the beam with respect to the neutral axis

$$M\,(max.) = \frac{Wl}{12} \qquad \ldots (14.5.33)$$

Where M is the maximum bending moment at the edges. Maximum vertical shear load V is given by,

$$V\,(max.) = \frac{W}{2} \qquad \ldots (14.5.34)$$

The maximum fibre stress σ in the beam occurs at the section of greatest bending moment and is expressed as,

$$\sigma\,(max.) = \frac{M}{I/c} \qquad \ldots (14.5.35)$$

Where c is the distance of extreme fibre from the neutral axis.

14.6 DESIGN OF PERFORATED SIEVE PLATE AND SUPPORT PLATE FOR PACKED BED

Design procedure is the same as that described in section 14.5. Due to perforations sieve plates get weaker. To take care of this an efficiency factor is to be introduced. If t is the minimum thickness for unpierced plate and t_p the desired perforated plate thickness under same operating conditions, then,

DESIGN OF SOME SPECIAL PARTS

$$\phi = \frac{t}{t_p} = \frac{p_h - d_h}{p_h} \qquad \ldots (14.6.1)$$

Where,

$\phi =$ reduced strength factor for perforated plates

$p_h =$ pitch of perforations

$d_h =$ perforated hole diameter

Using appropriate equations from section 14.5, unpierced plate thickness t is to be determined. From the sieve plate layout, reduced strength factor ϕ can be determined. Dividing t by ϕ will give t_p.

CHAPTER 15

OPTIMIZATION

15.1 INTRODUCTION

To-day 'optimization' is not a new terminology to engineering students or practising engineers. Selecting the "best" answer to a problem out of the multiplicity of potential solutions is possible by optimization techniques. There are, however, a number of reasons why engineers are becoming greatly conscious about optimization. An important one is that intensive competition in the chemical process industries makes it more necessary than ever that equipment and systems operate at maximum efficiency. Moreover, the complex process conditions for the production of many modern chemicals specify very costly materials like austenitic stainless steel, monel, titanium, etc. Savings of these materials even in small quantity can greatly help in improving the economy.

In this chapter firstly, a few widely used optimization techniques will be discussed and next their applications in equipment design will be shown. It is assumed that some basic concepts of maxima, minima, tangents and normals that are important for an understanding of optimization are already possessed by the readers.

Before adopting a particular optimization technique one is to first decide "what to optimize". For this an "objective function" is to be selected first. This may either be related to cost, material consumption, surface area or some other criterion. Next is to determine the parameters which may influence the objective functions. Finally a relationship is to be derived between objective function and the influencing parameters. These are the preamble steps to be completed before asking "how to optimize".

The various methods of optimization may be classified as follows:
(a) Analytical
(b) Case study
(c) Search

(a) Analytical Method

The classical method of calculus is used in the analytical method when the first derivative of a differentiable function is set equal to zero.

In the case of a multivariable function, partial derivatives are taken with respect to each of the n variables, and a set of n simultaneous equations is obtained. Introducing

Lagrange Multipliers the number of independent equations and the number of unknowns are made identical, and the solution is obtained.

A serious limitation regarding the application of analytical method is that the function must be in a mathematical form and must be differentiable. Many design problems are of the iterative type, solved by trial and error or some numerical method. Analytical method gives only one solution and it does not give any indication of the direction or path to reach the solution.

(b) Case Study Method

As the name implies, this technique consists of evaluating various solutions to the problem. Case after case is tried and results are presented graphically or in tabular form. Analysing these graphs and tables the "best" combination of variables is arrived at.

(c) Search Methods

Various search methods are reported in the literature[1-3]. They have certain characteristics in common. A base point in all cases is assumed known, that means, one set of conditions as a solution to the problem, but not necessarily the optimum solution. Then, the method must select the next set of values for the variables and evaluate the objective function once again, hoping that this time, and similarly each successive time, the solution will be closer to the optimum solution.

Each search method, though, has this objective in common, each selects the next set of values in a different manner. One requirement, however, in all these methods is that there should be only one optimum solution.

15.2 OPTIMIZATION TECHNIQUES

In the following sections a few widely used techniques will be elaborated indicating the methods of their applications in optimizing chemical equipment.

15.2.1 Functions of one variable system

Rules about extremes of a function of one variable may be summarized as follows[4] :

1. Extremes of a function can occur only where the first derivative becomes zero to non-existent.

2. If second derivative of the function is negative, a maximum exists ; if it is positive, a minimum exists.

---

[1] H. Hancock, "Theory of Maxima and Minima", Dover Pub., New York, 1960.

[2] D. J. Wilde, "Optimum Seeking Methods", Prentice-Hall, Englewood Cliffs, N.J., 1964.

[3] D. F. Rudd and C. C. Watson, "Strategy of Processing Engineering", John Wiley and Sons, Inc., New York, 1968.

[4] A. H. Boas, "Optimization", Chemical Engineering, December 10, 1962.

OPTIMIZATION

Design Example 15.1: It is required to determine the optimum diameter to height ratio for a large oil storage vessel, so that the total cost is minimum. Following data may be used for calculation.

V = volume of the vessel
= 1 000 m³

C_s = cost of the sides per sq. meter

C_h = cost of the head or top per sq. meter
= 1.5 C_s

C_b = cost of the bottom per sq. meter
= 0.75 C_s

Solution:
Let, C_T = total cost of the vessel
D = diameter, m
H = height, m

Then,

$$C_T = C_s \pi D H + (C_b + C_h) \frac{\pi D^2}{4} \quad \ldots (1)$$

$$V = \frac{\pi D^2 H}{4} \quad \ldots (2)$$

or, $H = \dfrac{4V}{\pi D^2}$

Substituting for H in (1)

$$C_T = C_s \cdot 4 \frac{V}{D} + (C_b + C_h) \frac{\pi D^2}{4} \quad \ldots (3)$$

Differentiating with respect to design variable D and equating to zero

$$\frac{dC_T}{dD} = -\frac{4 C_s V}{D^2} + (C_b + C_h) \frac{\pi D}{2} = 0$$

or, $$D^3 = \frac{C_s}{C_b + C_h} \cdot \frac{8V}{\pi} \quad \ldots (4)$$

$$\frac{d^2 C_T}{dD^2} = \frac{8 C_s V}{D^3} = \text{positive}$$

Substituting for the volume in (4) in terms of D and H gives the following optimum D/H ratio for the minimum cost of the vessel.

$$\frac{D}{H} = \frac{2 C_s}{C_b + C_h} \quad \ldots (5)$$

Substituting for C_b and C_h in terms of C_s gives.

$$\frac{D}{H} = \frac{2 C_s}{(0.75 + 1.5) C_s} = \frac{8}{9}$$

It appears from relation (5) the optimum D/H ratios depend upon the unit costs of various parts.

15.2.2 Lagrange multipliers

In earlier sections it has been mentioned that the analytical method of optimization involves setting the derivatives of functions equal to zero. One extremely powerful analytical technique — use of Lagrange Multipliers — is applied when there are equality constraints or restrictions on the variables (e.g. purity must equal a certain value, flow rate must equal a design value).

The object of optimization studies is to determine values of the independent variables that will maximize or minimize some objective function (e.g. cost as a function of diameter and height of a vessel). A Lagrange Expression is developed first. Then, values of the independent variables that optimize this expression are determined, subject to the given constraints of the problem. The objective function will also be optimized by these same values of the independent variables.

There are two important rules in applying this technique :

1. The number of Lagrange Multipliers to be introduced must be equal to the number of constraining equation ;
2. The Lagrange Expression must be equal to the objective function plus the product of the Lagrange Multiplier and constraint. This constraint must be in the form of an equation set equal to zero.

Development of Lagrange Expression :

Let, $u(x, y, \ldots\ldots)$ = objective function

$v_1(x, y, \ldots\ldots) = 0$ and $v_2(x, y, \ldots\ldots) = 0$
etc. are constraining equations.

λ_1, λ_2, etc. are Lagrange Multipliers

$w(x, y, \ldots\ldots)$ = Lagrange Expression.

$x, y, \ldots\ldots$ are independent variables.

Then, according to the definition of the Lagrange Expression

$$w(x, y, \ldots\ldots) = u(x, y, \ldots\ldots) + \lambda_1 v_1(x, y, \ldots\ldots) + \lambda_2 v_2(x, y, \ldots\ldots) \quad \ldots (15.2.1)$$

Once Lagrange Expression is developed as shown by Eq. 15.2.1, next step will be to take partial derivatives of the Lagrange Expression with respect to each independent variable (including the Lagrange Multipliers) and each must be set equal to zero for an extreme. Thus N simultaneous equations are obtained, one for each of the N independent variables, and it is possible to solve for maximum or minimum conditions. Since each constraining equation introduces one additional equation, one Lagrange Multiplier for each equation is introduced to compensate for this.

Design Example 15.2 : What will be dimensions of an open the rectangular tank of $1\,000$ m³ capacity to give the minimum area.

Solution :

Let, V = volume of the tank
 = $1\,000$ m³
H = height of the tank, m
L = length of the tank, m
B = breadth of the tank, m

Then, $\quad V = LBH$

The objective function is the area to be minimized and it is equal to

$$u(L, B, H) = 2HB + 2HL + LB$$

The constraining equation is

$$v(L, B, H) = LBH - V$$

The Lagrange Multiplier is λ and the Lagrange Expression is

$$w(L, B, H) = 2HB + 2HL + LB + \lambda(LBH - V)$$

Taking partial derivatives with respect to L, B, H and λ, and setting the resulting equations equal to zero,

$$\frac{\partial w}{\partial L} = 2H + B + \lambda BH = 0 \qquad \ldots (1)$$

$$\frac{\partial w}{\partial B} = 2H + L + \lambda LH = 0 \qquad \ldots (2)$$

$$\frac{\partial w}{\partial H} = 2B + 2L + \lambda LB = 0 \qquad \ldots (3)$$

$$\frac{\partial w}{\partial \lambda} = LBH - V = 0 \qquad \ldots (4)$$

From relation (1),

$$\lambda = -\frac{(B + 2H)}{BH} \qquad \ldots (5)$$

Substituting in (2) and solving for B,

$$B = L \qquad \ldots (6)$$

From (3), it is found that

$$\lambda = -\frac{4}{B} \qquad \ldots (7)$$

From (1),

$$B = 2H \qquad \ldots (8)$$

254 CHEMICAL EQUIPMENT DESIGN—MECHANICAL ASPECTS

Therefore,
$$V = LBH = (B)(B)(B/2)$$
$$= \frac{B^3}{2}$$
$$= 1\,000 \text{ m}^3$$

From this
$$B = 12.6 \text{ m}$$
$$L = 12.6 \text{ m}$$
$$H = 6.3 \text{ m}$$

Design Example 15.3 : One of the chemical manufacturers stores polyethylene pellets in 50 m³ cylindrical vessels with 60° conical bottoms. The actual size is some 2 7 m diameter and 8.0 m cylinder height. Assuming that the per area cost of the flat top and conical bottom is n times that of the cylindrical sides C_s, derive an expression for the minimum cost dimensions for such vessels. In practice, n is on the order of two ; how does the theory compare to the practice ?

Solution :

If V = volume of a vessel, m³
H = cylindrical height, m
D = diameter, m
α = half the apex angle of the conical bottom, deg.

Then, $V = \dfrac{\pi D^2}{4} H + \left(\dfrac{1}{3}\right)\left(\dfrac{\pi D^2}{4}\right)\left(\dfrac{D}{2 \tan \alpha}\right)$

Substituting $\alpha = 30°$
$$V = \frac{\pi D^2}{4} H + \frac{\pi D^3}{13.8}$$

The objective function is the cost of the vessel to be minimum and it is equal to

$$u(D, H) = (\pi D H) C_s + \frac{\pi D^2}{4}(n C_s) + \left(\frac{\pi D}{2}\right)\left(\frac{D}{2 \sin \alpha}\right)(n C_s)$$

Substituting for $\alpha = 30°$
$$u(D, H) = \pi C_s (D H + 0.75 D^2 n)$$

The constraining equation is
$$v(D, H) = \frac{\pi D^2}{4} H + \frac{\pi D^3}{13.8} - V$$

The Lagrange Multiplier is λ and the Lagrange Expression is
$$w(D, H) = \pi C_s (D H + 0.75 D^2 n) + \lambda \left(\frac{\pi D^2}{4} H + \frac{\pi D^3}{13.8} - V\right)$$

Taking partial derivatives with respect to D, H and λ, and setting the resulting equations equal to zero,

$$\frac{\partial w}{\partial D} = \pi C_s (H + 1.5\, Dn) + \lambda \left(\frac{\pi DH}{2} + \frac{\pi D^2}{4.6} \right) = 0 \qquad \ldots (1)$$

$$\frac{\partial w}{\partial H} = \pi C_s (D) + \lambda \left(\frac{\pi D^2}{4} \right) = 0 \qquad \ldots (2)$$

$$\frac{\partial w}{\partial \lambda} = \frac{\pi D^2}{4} H + \frac{\pi D^3}{13.8} - V = 0 \qquad \ldots (3)$$

From equation (2), $\lambda = -\dfrac{4 C_s}{D}$

Substituting for λ in equation (1) and eliminating πC_s,

$$H + 1.5\, Dn - 2H - 0.87\, D = 0$$

or, $\quad H/D = 1.5\, n - 0.87$

if $\quad n = 2,\ H/D = 2.13$

Substituting for H in equation (3),

$$1.898\, D^3 = V = 50$$

or $\quad D = 2.98$ m

and $\quad H = 2.13\, D$

$\quad\quad\quad = 6.70$ m

Compared to the economical $H/D = 2.13$, in practice $H/D = \dfrac{8}{2.7} = 2.95$ is taken.

15.2.3 Search by Golden Section

Depending upon the method of selecting the search points number of calculations required to reach the optimum or near the optimum value will depend. Search by the Golden Section is an efficient technique to arrive at the optimum value[3]. In this method search points are symmetrically placed from both the ends of search region. Now the question is how to determine the symmetrical search points ?

Suppose that two search points are placed at a distance 'l' from each end of the search region. On the basis of these two calculations, a portion of the region can be eliminated, where optimum value cannot lie. This region elimination is done by examining the two results. The system considered here is unimodal, i.e., it will have only one peak point. Let A and B are two ends of the search region varying 0 to 1. Within 0 to 1 it is to be determined where maximum is lying. Let C and D are two search points within the specified region taken at a distance 'l' from A and B respectively, and found that search value at C is higher than at D. According to the conception of unimodality or one peak point, it is clear that maximum cannot lie in the region A to D and therefore this region is eliminated for further search. Now the search region is reduced to D to B which is equal to 'l' fraction of the original region. The point C is at a distance $(1 - l)$ from B.

Length of new search region is now 'l'. Therefore, the distance of the new search points would be l^2 from D and B within the region. The characteristic of 'l' is such that l^2 from B is nothing but C, which is again equal to $(1 - l)$.

This leads to the following relationship:

$$l^2 = 1 - l$$

or, $$l = \frac{\sqrt{5} - 1}{2} = 0.618.$$

In this way, by placing the first two points a distance of 0.618 of the original region of isolation from the sides, symmetry will be preserved. Each new point reduced the region of isolation by 0.18. Following table gives the efficiency of the method.

Step Number	New Calculations	Total Calculations	Region of Isolation of *optimum*
1	2	2	(0.618)
2	1	3	$(0.618)^2$
3	1	4	$(0.618)^3$
.			
.			
.			
m	1	$N = 1 + m$	$I_R = (0.618)^m$

In the above table N is the total number of calculations and I_R region of isolation. From this,

$$I_R = (0.618)^{N-1} \quad \text{and} \quad N = 1 + \frac{\log I_R}{\log 0.618}$$

Design Example 15.4: A storage tank for 50 per cent NaOH solution of relative density = 1.525 at 20 °C is to be designed. The bottom rests on a concrete pad, is 6 mm plate, and will cost (complete with the pad) Rs. 750 per square meter. The top is a 3 mm supported roof, costing Rs. 300 per square meter. The thickness, t mm, of the cylindrical sides will depend upon the hydrostatic pressure, with a minimum of 5 mm as follows:

$$t \geqslant 5 \text{ mm}$$

$$t = \frac{pD}{2f} = \frac{pD}{170}$$

for $f = 85$ MN/m^2

p = pressure, MN/m^2

D = tank diameter, mm

The sides cost 75 t rupees per square meter.

OPTIMIZATION

Using the search by the Golden Section, determine the geometry and cost of the optimum tank to store:

(a) 30 m³ of NaOH
(b) 3 000 m³ of NaOH.

Solution : This problem could be solved by Calculus or Lagrange Multipliers. Here application of Golden Section method will be shown.

For, V = volume of the tank, m³
D = diameter of the tank, mm
H = height of the tank, mm

$$V \times 10^9 = \frac{\pi D^2}{4} H \qquad \ldots (1)$$

$$p = \frac{H}{1\,000} \gamma \qquad \ldots (2)$$

Where, γ = specific weight of NaOH solution
$= 1.525 \times 10^3 \times 9.807 \times 10^{-6}$ MN/m³
$= 1.51 \times 10^{-2}$ MN/m³

Substituting in equation (2),

$$p = 1.51 \times 10^{-5} H \qquad \ldots (2a)$$

$$t = \frac{pD}{170}$$

Substituting for p,

$$t = 9 \times 10^{-8} HD \qquad \ldots (3)$$

From equation (1),

$$HD = \frac{1.27 \times 10^9 \, V}{D} \qquad \ldots (4)$$

Substituting in (3)

$$t = 114.3 \frac{V}{D} \qquad \ldots (5)$$

$$D = 114.3 \frac{V}{t} \qquad \ldots (6)$$

$$D^2 = 1.36 \times 10^4 \frac{V^2}{t^2} \qquad \ldots (7)$$

$$DH = 1.1 \times 10^7 \, t \qquad \ldots (8)$$

Let, C_T = total cost of the tank, Rs.
C_b = cost of bottom per m², Rs.
$= 750$

D—33

c_t = cost of top per m², Rs.
= 300

c_s = cost of sides per m², Rs.
= 75 t

$$C_T = \frac{\pi D^2}{1 \times 10^6}(c_b + c_t) + \frac{\pi DH}{10^6}(c_s) \quad \ldots (9)$$

Substituting and simplifying,

$$C_T = 11.2\, V^2/t^2 + 2\,600\, t^2 \quad \ldots (10)$$

Value of t is to be determined for which C_T will be minimum.

(a) For $V = 30$ m³,

$$C_T = \frac{1 \times 10^4}{t^2} + 2\,600\, t^2$$

In Golden section method, the selection of region of search is very important. Depending on this, number of search can be minimized to arrive at the optimum.

For $V = 30$ m³, let it be assumed that region of search is defined by $t = 0$ to 10 mm.

t (mm)	C_T (Rs)	*Remarks*
6	93 778 ⎱	t must be less than 6 mm.
4	42 225 ⎰	Again 4 mm does not satisfy the condition $t >$ 5 mm. So, t must be larger than 4 mm.
5	65 403	$t = 5$ mm is to be selected.

From equation (6)

$$D = 114.3\,\frac{V}{t}$$
$$= 114.3\,\frac{30}{5}$$
$$= 685.8 \text{ mm}$$

From equation (8)

$$H = 1.1 \times 10^7\,\frac{t}{D}$$
$$= \frac{1.1 \times 10^7 \times 5}{685.8}$$
$$= 8 \times 10^4 \text{ mm}$$
$$H/D = \frac{8 \times 10^4}{685.8} = 117$$

It may be noted that H/D ratio is so high that the dimensions of the tank are not practical.

Therefore, the situation is to be re-examined. From equation (3) it can be said that $t = 5$ is valid for any value of H and D. If (HD) does not exceed 5.56×10^7. In this context equation (9) is to be further analysed. Here, C_s can also be taken constant and equal to (75×5) or 375.

Substituting for H in equation (9) from equation (1)

$$C_T = 7.95 \times 10^{-4} D^2 + \frac{4.5 \times 10^7}{D} \qquad \ldots (11)$$

Assume region of search for $D = 0$ to 5 000 mm.

D (mm)	C_T (Rs.)	Remarks
1 900	26 570	Optimum value will be lying above $D = 1\,900$
3 100	22 120	
3 800	23 400	Optimum will be below $D = 3\,800$
2 600	22 700	Optimum will be above $D = 2\,600$, but below $D = 3\,800$
3 340	22 800	Optimum lies below $D = 3\,340$ and above $D = 2\,600$
2 900	22 200	Optimum lies between $D = 2\,900$ and 3 100

Before going for further calculations, its necessary is to be thought of, as cost difference is Rs. 80 only. Therefore a decision can be taken at this point in favour of $D = 3\,000$ which is a standard value.

For $D = 3\,000$; $C_T = 22\,155$

$H = 4\,250$

$H/D = 1.42$

$HD = 1.28 \times 10^7 < 5.56 \times 10^7$

In this example it has been seen that in optimizing any system, selection of controlling parameter is more important. In this case t is not a controlling parameter.

(b) For $V = 3\,000$ m³, from equation (10)

$$C_T = \frac{1 \times 10^8}{t^2} + 2\,600\, t^2$$

Assume region of search for t from 10 mm to 20 mm for the first check. Following the similar procedure optimum value of t is found to be equal to 14 mm and $C_T = $ Rs. 1.02×10^6. Tank dimensions are :

$D = 24\,000$ mm

$H = 6\,600$ mm

$H/D = 0.275$

For large storage tanks, such dimension are reasonable. Height could be a little more. But considering the effect of wind load on large diameter tanks, small H/D ratio is usually recommended.

APPENDICES

APPENDIX A — MECHANICAL PROPERTIES OF METALS

Table A-1 Mechanical Properties of Carbon and Low Alloy Steels

Material specification	Grade	U.T.S. 10^8 N/m^2	Yield stress 10^8 N/m^2	% Elongation	Allowable stress values in 10^8 N/m^2 or temperature upto °C								
					250	300	350	400	450	475	500	525	550
PLATES													
IS : 2002-1962	1	3.63	1.98	26	0.93	0.85	0.76	0.70	0.42	0.35	...	...	...
	2 A	4.12	2.06	25	0.96	0.88	0.79	0.72	0.42	0.35	...	...	...
	2 B	5.10	2.55	20	1.18	1.08	0.98	0.81	0.42	0.35	...	...	...
IS : 2041-1962	20 Mo 55	4.71	2.75	20	1.40	1.29	1.20	1.12	1.05	0.75	0.54	0.36	...
	20 Mn 2	5.10	2.95	20	1.37	1.25	1.13	0.81	0.42	0.35	...	...	...
IS : 1570-1961	15 Cr 90 Mo 55	4.91	2.95	20	1.56	1.49	1.41	1.31	1.23	1.14	0.84	0.56	0.34
	C 15 Mn 75	4.12	2.26	25	1.04	0.96	0.87	0.79	0.42	0.35	...	...	...
FORGINGS													
IS : 2004-1962	Class 1	3.63	1.82	...	0.84	0.77	0.69	0.63	0.42	0.35	...	...	...
	Class 2	4.32	2.16	15	1.00	0.91	0.83	0.75	0.42	0.35	...	...	...
	Class 3	4.91	2.46	21	1.14	1.04	0.94	0.81	0.42	0.35	...	...	...
	Class 4	6.18	3.09	15	1.44	1.31	1.19	0.81	0.42	0.35	...	...	...
IS : 1570-1961	20 Mo 55	4.71	2.75	20	1.40	1.29	1.20	1.12	1.05	0.75	0.54	0.36	...
	10 Cr 2 Mo 1	4.91	3.14	20	1.75	1.69	1.60	1.54	1.43	1.24	0.94	0.68	0.48
IS : 2611-1964	15 Cr 90 Mo 55	4.91	2.95	20	1.56	1.49	1.41	1.31	1.23	1.14	0.84	0.56	0.34

Table A-1 Continued

Table A-1 Continued

Material specification	Grade	U.T.S. 10^8 N/m²	Yield stress 10^8 N/m²	% Elongation	\multicolumn{9}{c}{Allowable stress values in 10^8 N/m² for temperature upto °C}								
					250	300	350	400	450	475	500	525	550
\multicolumn{14}{c}{TUBES, PIPES}
IS: 3609-1966	1% Cr ½% Mo	4.32	2.34	22	1.25	1.18	1.12	1.04	0.98	0.95	0.84	0.56	0.34
	2¼% Cr 1% Mo	4.81	2.46	19	1.37	1.32	1.25	1.20	1.13	1.10	0.94	0.68	0.48
IS: 1570-1961	20 Mo 55	4.51	2.46	21	1.25	1.15	1.07	1.01	0.94	0.75	0.54	0.36	...
IS: 1914-1961	U.T.S.	3.14	1.57	30	0.72	0.66	0.60	0.54	0.42	0.35	...	...	...
	U.T.S.	4.22	2.11	22	0.98	0.90	0.81	0.74	0.42	0.35	...	...	...
IS: 2416-1963	U.T.S.	3.14	1.57	30	0.72	0.66	0.60	0.54	0.42	0.35	...	...	...
IS: 1978-1961	St 18	3.10	1.73	...	0.80	0.73	0.65	0.60	0.42	0.35	...	...	...
	St 20	3.31	1.94	...	0.90	0.82	0.74	0.67	0.42	0.35	...	...	...
	St 21	3.31	2.07	...	0.96	0.88	0.79	0.72	0.42	0.35	...	...	...
	St 25	4.14	2.42	...	1.12	1.02	0.93	0.81	0.42	0.35	...	...	...
IS: 1979-1961	St 30	4.14	2.90	...	1.35	1.23	1.12	0.81	0.42	0.35	...	...	...
	St 32	4.35	3.17	...	1.47	1.35	1.22	0.81	0.42	0.35	...	...	...
	St 37	4.55	3.59	...	1.67	1.52	1.38	0.81	0.42	0.35	...	...	...

APPENDICES

Sections, Plates, Bars

IS : 226-1962	St 42-S	4.12	2.36	23	0.96	0.88	0.80	..	..
IS : 961-1962	St 55 HTW	4.92	2.84	20	1.12	1.05	0.94	..	..
IS : 2062-1962	St 42-W	4.12	2.25	23	0.96	0.88	0.80	..	..
IS : 3039-1965	Grade A	..	..	..	0.96	0.88	0.80	..	..
	Grade D	..	..	..	1.12	1.05	0.94	..	..
IS : 3503-1966	Grade 1	3.63	2.00	26	0.84	0.77	0.70	0.64	0.42	0.35
	Grade 2	4.12	2.27	25	0.96	0.88	0.80	0.72	0.42	0.35
	Grade 3	4.32	2.38	23	1.00	0.91	0.83	0.75	0.42	0.35
	Grade 4	4.61	2.54	22	1.12	1.05	0.94	0.81	0.42	0.35
	Grade 5	4.92	2.71	21	1.19	1.09	0.98	0.81	0.42	0.35
IS : 3945-1966	Grade A-N	4.32	2.36	23	0.96	0.88	0.80	..	..
	Grade B-N	4.92	2.80	20	1.12	1.05	0.94	..	..

Table A-2 Mechanical Properties of High Alloy or Austenitic Stainless Steels

Product : Plates, Sections, Bars, Foregins, Seamless tubes

Material specification	Designation	U.T.S. 10^8 N/m²	Yield stress 10^8 N/m²	% Elongation	Allowable stress values in 10^8 N/m² for temperature upto °C							
					50	100	150	200	250	300	350	400
IS : 1570-1961	04 Cr 19 Ni 9	5.40	2.75	28	1.57	1.39	1.23	1.04	0.98	0.92	0.85	0.79
	04 Cr 19 Ni 9 Ti 20	5.40	2.75	28	1.57	1.40	1.23	1.06	1.04	1.04	1.04	1.04
	04 Cr 19 Ni 9 Nb 40	5.40	2.75	28	1.57	1.40	1.23	1.06	1.04	1.04	1.04	1.04
	05 Cr 18 Ni 11 Mo 3	5.40	2.75	28	1.57	1.42	1.27	1.13	1.10	1.10	1.10	1.09
	05 Cr 19 Ni 9 Mo 3 Ti 20	5.40	2.75	28	1.57	1.42	1.27	1.13	1.10	1.10	1.10	1.09

Table A-3 Mechanical Properties of Copper and Copper Alloys

Material specification	Grade	U.T.S. 10^8 N/m²	% Elongation	Allowable stress values in 10^7 N/m² for temperature upto °C								
				50	75	100	125	150	175	200	225	250
PLATE, SHEET, STRIP												
IS: 410-1967	Cu Zn 30	2.75	45	6.89	6.89	6.89	6.89	6.82	5.59	3.75	2.41	..
	Cu Zn 37	2.75	45	8.62	8.50	8.13	7.65	7.13	5.27	1.96	..	..
	Cu Zn 40	2.75	30	8.62	8.50	8.13	7.75	7.13	5.27	1.96	..	..
IS: 1972-1961	All grades	2.21	35	4.61	4.57	4.45	4.21	3.40	2.65	1.86	..	..
TUBES												
IS: 407-1966	Alloy 1	2.85	..	6.89	6.89	6.89	6.89	6.82	5.59	3.75	..	..
	Alloy 2	2.85	..	8.62	8.50	8.13	7.65	7.13	5.27	1.96	..	..
IS: 1545-1960	ISBT 1 ISBT 2	..	..	6.89	6.89	6.89	6.89	6.82	5.59	3.75	..	..
	ISABT	..	..	8.27	8.27	8.27	8.27	8.12	5.32	2.53	1.54	..
	ISABZT	..	..	8.59	8.50	8.36	8.17	7.93	6.95	4.55	3.09	1.84
IS: 2371-1963	Cu Zn 2 1 Al 2 As	3.14	..	8.27	8.27	8.27	8.27	8.12	5.32	2.53	1.54	..
	Cu Ni 31 Mnl Fe	4.12	..	8.14	7.92	7.73	7.56	7.44	7.43	7.12	7.00	6.97
IS: 2501-1963	...	..	..	4.13	4.10	4.05	3.92	3.40	2.65	1.86	..	..

Table A-4 Values of E (Modulus of Elasticity) for Ferrous Materials in 10^{11} N/m²

Material	Design Temperature °C					
	0	20	100	200	300	400	500
Carbon steels							
C ≤ 0.3%	1.93	1.93	1.91	1.87	1.79	1.67	..
C > 0.3%	2.07	2.06	2.03	1.96	1.87	1.70	..
Carbon-molybdenum steels and chromemolybdenum steels (upto 3% Cr)	2.07	2.06	2.03	1.97	1.91	1.81	1.69
Intermediate chrmoe molybdenum steels and austenitic stainless steels	1.90	1.90	1.87	1.83	1.77	1.70	1.57

APPENDICES

Table A-5 Values of E (Modulus of Elasticity) for Nickel and Nickel Alloy in 10^{11} N/m²

Material	Design Temperature °C							
	200	300	400	500	600	700	750
Nickel	2.07	2.00	1.85	1.62	1.38	1.15	1.07
70% Nickel and 30% copper alloy	1.85	1.77	1.73	1.66	1.59	1.52	1.47
75% Nickel, 15% Chromium and 10% ferrous alloy	2.15	2.04	1.97	1.73	1.57	1.28	1.17

Table A-6 Values of E (Modulus of Elasticity) for Copper and its Alloys in 10^{11} N/m²

Material	Design Temperature °C								
	20	50	100	150	200	250	300	350
Copper (99.98% Cu)	1.10	1.09	1.08	1.06	1.04	1.02	0.99	0.95
Commercial brass (66% Cu, 34% Zn)	0.96	0.95	0.94	0.93	0.89	0.87	0.84	0.83
Leaded tin bronze (88% Cu, 6% Sn, 1.5% Pb, 4.5% Zn)	0.89	0.88	0.87	0.85	0.82	0.80	0.78	0.75
Phosphor bronze (85.5% Cu, 12.5% Sn, 10% Zn)	1.03	1.01	1.00	0.96	0.93	0.89	0.83	0.65
Muntz (59% Cu, 39% Zn)	1.05	1.00	0.96	0.89	0.81	0.75	...	...
Cupro nickels (80% Cu, 20% Ni or 70% Cu, 30% Ni)	1.31	1.29	1.27	1.24	1.22	1.19	1.16	1.13

Table A-7 Mean Coefficient of Thermal Expansion in 10^{-5} m/m °C Between 20 °C and Specified Temperature

Material	Temperature °C														
	50	100	150	200	250	300	350	400	450	500	550	600	650	700	750
Carbon and C-Mo steels and low Cr steels upto 3% Cr	1.12	1.15	1.19	1.22	1.26	1.29	1.32	1.36	1.40	1.42	1.44	1.46	1.48	1.50	1.51
Intermediate alloy steels 5 Cr-Mo to 9 Cr-Mo	1.06	1.08	1.11	1.13	1.17	1.20	1.22	1.24	1.27	1.28	1.30	1.31	1.33	1.35	1.37
Straight chromium steels 12, 17, 25% Cr	0.97	0.99	1.02	1.04	1.08	1.10	1.12	1.14	1.16	1.17	1.19	1.21	1.22	1.24	1.24
Austenitic stainless steels	1.66	1.67	1.71	1.73	1.75	1.76	1.79	1.81	1.83	1.84	1.85	1.87	1.89	1.91	1.91
25 Cr-20 Ni steel	..	1.40	1.42	1.46	1.48	1.50	1.53	1.55	1.57	1.59	1.60	1.62	1.64	1.64	1.66
Grey cast iron	1.02	1.04	1.08	1.10	1.13	1.16	1.20	1.21	1.24	1.27	1.29	..	..	..	..
Cupro-nickel, 70 Cu-30 Ni	..	1.53	1.57	1.60	..	..	..	..	..	..	..	..	..	..	..
Copper	1.70	1.73	1.75	1.77	1.78	1.81	1.83	1.85	1.86	1.88	1.89	..	..	..	..
Brass, 66 Cu-34 Zn	1.73	1.75	1.80	1.83	1.88	1.92	1.94	1.99	2.04	2.06	2.10	2.14	2.18	..	..
Nickel-copper (Monel)	1.31	1.33	1.37	1.40	1.44	1.46	1.48	1.50	1.52	1.54	1.55	1.57	..	..	..
Aluminium	2.30	2.34	2.39	2.44	2.50	2.54	..	..	..	..	..	..	..	..	..
Ni-Cr-Fe (Inconel)	1.16	1.19	1.22	1.26	1.30	1.34	1.38	1.40	1.44	1.47	1.50	1.53	1.55	1.57	1.60
Nickel	1.28	1.30	1.35	1.39	1.40	1.43	1.47	1.43	1.51	1.54	1.55	1.57	1.58	1.60	1.60
Admiralty metal	..	..	..	..	..	2.02	..	..	..	..	..	..	..	..	..

Table A-8 Values of Specific Weight and Poisson's Ratio of Some Metals

Material	Specific weight N/m³	Poisson's ratio
Aluminium	2.65×10^4	0.34
Brass	8.35×10^4	0.35
Copper	8.79×10^4	0.35
Iron	7.74×10^4	0.28
Nickel	8.74×10^4	0.36
Steel	7.70×10^4	0.30

REFERENCES:

1. IS : 2825 — 1969 Code for Unfired Pressure Vessels.
2. IS : 4503 — 1969 Specification for Shell and Tube Type Heat Exchangers.

APPENDIX B — STANDARD VALUES

Table B-1 Steel Plates

Thickness (mm): 5, 5.5, 6, 7, 8, 9, 10, 11, 12, 14, 16, 18, 20, 22, 25, 28, 32, 36, 40, 45, 50, 56, 63, 71, 80.

Width (mm): 160, 180, 200, 220, 250, 280, 320, 355, 400, 450, 500, 560, 630, 710, 800, 900, 1 000, 1 100, 1 250, 1 500, 1 600, 1 800, 2 000, 2 200, 2 500.

Length: 6 m — 10 m easily available.

Table B-2 Strip Steels

Thickness (mm): 0.8, 0.9 1.0, 1.1, 1.2, 1.4, 1.6, 1.8, 2.0, 2.2, 2.5 2.8, 3.2, 3.5, 4.0, 4.5.

Width (mm): 100, 110, 125, 140, 160, 180, 200, 220, 250, 280, 320, 355, 400, 450, 500, 560, 630, 710, 800, 900, 1 000.

Table B-3 System International Screw Threads (S.I.)

Size (mm)	Root dia. (mm)	Size (mm)	Root dia. (mm)
6	4.59	36	30.37
8	6.24	39	33.37
10	7.89	42	35.68
12	9.54	45	38.68
14	11.19	48	40.99
16	13.19	52	44.99
18	14.48	56	48.20
20	16.48	60	52.20
22	18.48	64	56.20
24	19.78	68	59.52
27	22.78	72	63.52
30	25.07	76	67.52
33	28.07	80	71.52

Table B-4 Nominal Diameter for Process Equipment

(Usually O.D. in mm)

100, 125, 150, 200, 250, 300, (350), 400, 500, 600, 700, 800, 900, 1 000, 1 100, 1 200, (1 300), 1 400, (1 500), 1 600, (1 700), 1 800, (1 900), 2 000, (2 100), 2 200, (2 300), 2 400, 2 600, 2 800, 3 000, 3 200, (3 400), 3 600, (3 800), 4 000, (4 250), 4 500, (4 750), 5 000.

Values given in the brackets are second preference values.

Table B-5 Steel Specifications

A. Steel plates for fusion welded pressure vessels (IS : 2041—1962).
Type 1 Steel 20 Mo 55 and Type 2 Steel 20 Mn 2

B. Plain carbon steel plates for boilers (IS : 2002—1962) :

Grade 1—Plates for fire boxes and boiler plates which are required to be either welded, flanged or flame-cut.

Grade 2A—Plates of non-flanging quality (low tensile)—weldable.

Grade 2B—Plate of non-flanging quality (high tensile)—special precautions are necessary for welding.

C. Structural steels :

IS : 226—1962 (St 42-S)—Standard quality (tested steel) for sections, plates and bars.

IS : 2062—1962 (St 42-W)—Fusion welding quality (tested steel) for sections, plates and bars. Costlier than St 42-S.

CHEMICAL EQUIPMENT DESIGN—MECHANICAL ASPECTS

Table B-6 Calculation Aid for Determining Shell Diameter for Single-Pass Shell and Tube Heat Exchanger

D_1	n	n_1	D_1	n	n_1	D_1	n	n_1	D_1	n	n_1
0.000 0	1	1	21.633 4	433	12	32.741 4	967	12	40.841 2	1 519	12
2.000 0	7	6	22.000 0	439	6	32.924 2	979	12	41.036 6	1 531	12
3.464 2	13	6	22.271 0	451	12	33.045 4	1 003	24	41.328 0	1 555	24
4.000 0	19	6	22.538 8	463	12	33.286 6	1 015	12	41.569 2	1 561	6
5.291 6	31	12	22.715 6	475	12	33.406 6	1 027	12	41.717 4	1 573	12
6.000 0	37	6	23.065 2	499	24	33.644 2	1 039	12	41.761 2	1 585	12
6.928 2	43	6	23.579 6	511	12	34.000 0	1 045	6	41.904 6	1 597	12
7.211 2	55	12	24.000 0	517	6	34.117 4	1 057	12	42.000 0	1 615	18
8.000 0	61	6	24.248 8	535	18	34.176 0	1 069	12	42.142 6	1 627	12
8.717 8	73	12	24.331 0	547	12	34.641 0	1 075	6	42.332 0	1 639	12
9.163 2	85	12	24.576 4	559	12	34.698 8	1 099	24	42.567 6	1 651	12
10.000 0	91	6	24.980 0	571	12	34.871 2	1 111	12	42.755 2	1 663	12
10.392 3	97	6	25.060 0	583	12	35.042 8	1 123	12	42.034 8	1 675	12
10.583 0	109	12	25.534 3	595	12	35.156 8	1 135	12	43.266 6	1 687	12
11.135 6	121	12	26.000 0	613	18	35.383 6	1 146	12	43.312 8	1 711	24
12.000 0	127	6	26.153 4	625	12	35.552 8	1 159	12	43.405 0	1 723	12
12.165 6	139	12	26.229 8	637	12	36.000 0	1 165	6	43.589 0	1 735	12
12.490 0	151	12	26.457 6	649	12	36.055 6	1 177	12	43.863 4	1 759	24
13.114 8	163	12	26.907 2	661	12	36.166 2	1 189	12	44.000 0	1 765	6
13.856 4	169	6	27.055 4	673	12	36.386 8	1 201	12	44.136 2	1 777	12
14.000 0	187	18	27.495 4	685	12	36.496 6	1 213	12	44.226 6	1 789	12
14.422 2	199	12	27.712 8	691	6	36.660 6	1 225	12	44.542 2	1 801	12
15.099 6	211	12	27.784 8	703	12	36.715 2	1 237	12	44.676 6	1 813	12
15.620 4	223	12	28.000 0	721	18	37.040 6	1 261	24	45.033 4	1 831	18

APPENDICES

15.874 6	235	12	28.213 4	733	12	37.363 0	12	45.177 8	12
16.000 0	241	6	28.354 8	745	12	37.470 0	12	45.210 6	24
16.370 7	253	12	28.844 4	755	12	38.000 0	18	45.291 9	12
17.088 0	265	12	29.051 6	769	12	38.105 2	6	45.431 2	12
17.320 6	271	6	29.461 8	793	24	38.157 6	24	45.738 4	12
17.435 6	283	12	29.597 2	805	12	38.314 4	12	45.825 8	12
17.776 4	295	12	29.866 4	817	12	38.574 6	12	46.000 0	6
18.000 0	301	6	30.000 0	823	6	38.626 4	12	46.130 2	24
18.330 4	313	12	30.199 4	835	12	38.935 8	12	46.518 8	12
19.078 8	337	24	30.265 4	847	12	39.038 4	12	46.604 8	12
19.287 4	349	12	30.799 6	859	12	39.344 6	12	46.776 0	12
19.697 8	361	12	31.048 4	871	12	39.395 4	12	46.861 4	12
20.000 0	367	6	31.177 0	877	6	39.849 8	12	47.032 0	24
20.297 8	379	12	31.241 0	889	12	39.950 0	24	47.159 4	12
20.784 6	385	6	31.432 4	913	24	40.000 0	6	47.286 4	24
20.880 6	397	12	31.749 0	925	12	40.099 8	24	47.623 6	12
21.071 4	409	12	32.000 0	931	6	40.447 4	12	47.791 2	12
21.166 0	421	12	32.187 0	955	24	40.595 6	12	48.000 0	6

D_1 = outermost tube-circle diameter for pitch = 1 mm ;

n = total number of tubes or pipes ;

n_1 = number of the tubes lying on the outermost tube-circle ;

If, D = inside shell diameter, mm ;

d = outside tube diameter, mm ;

p_t = tube-pitch, mm ;

l_c = minimum clearance between outermost tubes and shell, mm ;

Then, $D = D_1$ (corresponding to n) $\times p_t + d + 2 l_c$.

Table B-7 Pressure Classification (IS : 4503—1967)

Heat exchangers are designated by the following design pressure at a basic metal temperature which is 250 °C, 120 °C and 65 °C respectively for carbon steel, stainless steel and non-ferrous metals :

2.5 bar, 6.3 bar, 10 bar, 16 bar, 25 bar and 40 bar

Table B-8(a) Minimum Shell Thickness for Petroleum Industry and Where Severe Conditions Expected

Nominal Dia. (mm)	Minimum Thickness (without corrosion allowance), (mm)		
	Carbon steel		
	Pipe	Plate	Alloy
Up to and including 300	6.3	..	3.2
Over 300 up to and including 700	10	10	5
Over 700 up to and including 1 000	..	11.2	6.3
Over 1 000 up to and including 1 500	..	14	8

Table B-8(b) Minimum Shell Thickness under Ordinary Conditions

Nominal Dia. (mm)	Cast Iron	Carbon steel (Including corrosion allowance)	Copper and copper alloys	Austenitic Stainless steel	Monel Inconel
150	10	5	3.2	3.2	3.2
200	10	6.3	3.2	3.2	3.2
250	10	6.3	3.2	3.2	3.2
300	13	6.3	3.2	3.2	3.2
350	13	6.3	5	3.2	3.2
400	13	6.3	5	3.2	3.2
500	13	8	6.3	3.2	3.2
600	16	8	6.3	5	5
700	16	10	8.0	5	5
800	16	10	10	6.3	6.3
900	19	10	10	6.3	6.3
1 000	19	10	11.2	6.3	6 3
1 100	22	11.2	11.2	6.3	6.3

Table B-9 Tube Lengths

unless otherwise specified, overall straight lengths are :
0.5, 1.0, 2.5, 3.0, 4.0, 5.0, 6.0 meters

Table B-10 Tube Diameter and Thicknesses

Dimensions are presented as O.D. × (minimum thickness) in mm.

Carbon steel (Petroleum)	12 × 1.6; 16 × 1.6; 19 × 1.6; 19 × 2.0; 25.4 × 2.0; 25.4 × 2.6; 25.4 × 3.2; 31.8 × 2.0; 3.18 × 2.6; 31.8 × 3.2; 38 × 3.2.
Carbon steel (Ordinary Industry)	19 × (1.6, 2.0); 25.4 × (2.0, 2.6); 31.8 × (2.0, 2.6, 3.2); 38 × 3.2.
Copper and its alloys (Petroleum)	(18, 20) × (1.2, 1.6, 2.0); 25 × (1.6, 2.0, 2.5); 32 × (2.0, 2.5, 3.0).
Copper and its alloys (Ordinary Industry).	6 × (0.6, 0.8); 10 × (0.8, 1.0, 1.2); 12 × (1.0, 1.2); 16 × (1.2, 1.6); (18, 20) × (1.2, 1.5, 2.0); 25 × (1.5, 2.0, 2.5); 32 × (2.0, 2.5, 3.0); 40 × 3.0.
Other alloys (Petroleum)	(18, 20) × (1.2, 1.5, 2.0); 25 × (1.2, 1.6, 2.0, 2.5); 32 × (1.6, 2.0, 2.5, 3.0).
Other alloys (Ordinary Industry)	6 × (0.6, 0.8); 10 × (0.8, 1.0, 1.2); 12 × (1.0, 1.2); (18, 20) × (1.2, 1.5, 2.0); 25 × (1.2, 1.6, 2.0, 2.5); 32 × (1.5, 2.0, 2.5, 3.0).

Table B-11 Minimum Tube Sheet Thickness (IS : 4503 — 1967)

Tube outer diameter (mm)	*Thickness* (mm)
6	6
10	10
12	12
16	13
18, 19, 20	15
25, 25.4	19
31.8, 32	22.4
38, 40	25.4

REFERENCES :
1. Soor's Diary by G. B. Kanuga.
2. IS : 4503 — 1967 Specification for Shell and Tube Type Heat Exchangers.
3. IS : 2825 — 1969 Code for Unfired Pressure Vessels.

APPENDIX C—PROPERTIES OF SECTIONS

Table C-1 Properties of Sections (Fig. C-1)

Section	Area of Section A	Distance from axis to extremities y and y_1	Moment of inertia I
1. Square	a^2	$y = a/2$	$\dfrac{a^4}{12}$
2. Square	a^2	$y = a$	$\dfrac{a^4}{3}$
3. Rectangle	bh	$y = \dfrac{h}{2}$	$\dfrac{bh^3}{12}$
4. Rectangle	bh	$y = h$	$\dfrac{bh^3}{3}$
5. Triangle	$bh/2$	$y = 2h/3$; $y_1 = \dfrac{h}{3}$	$\dfrac{bh^3}{36}$
6. Triangle	$bh/2$	$y = h$	$\dfrac{bh^3}{12}$
7. Trapezoid	$h(b + b_1)/2$	$y = \dfrac{h(b_1 + 2b)}{3(b_1 + b)}$; $y_1 = \dfrac{h(b + 2b_1)}{3(b_1 + b)}$	$\dfrac{h^3(b^2 + 4b\,b_1 + b_1^2)}{36(b + b_1)}$
8. Regular Hexagon	$0.866\,h^2$	$y = \dfrac{h}{2}$	$0.06\,h^4$
9. Regular Octagon	$0.828\,h^2$	$y = \dfrac{h}{2}$	$0.055\,h^4$

10. Solid circle	$0.785\, d^2$	$y = \dfrac{d}{2}$	$\dfrac{\pi d^4}{64}$
11. Hollow circle	$0.785\,(d^2 - d_1^2)$	$y = \dfrac{d}{2}$	$\dfrac{\pi (d^4 - d_1^4)}{64}$
12. Solid Semicircle	$\pi d^2/8$	$y = 0.288\, d;\ y_1 = 0.212\, d$	$0.006\,9\, d^4$
13. Hollow Semicircle	$\dfrac{\pi (d^2 - d_1^2)}{8}$	$y = 2\,(d^3 - d_1^3)/3\, \pi\, (d^2 - d_1^2)$ $y_1 = \dfrac{3\,\pi\, d\,(d^2 - d_1^2) - 4\,(d^3 - d_1^3)}{6\,\pi\,(d^2 - d_1^2)}$	$\dfrac{9\,\pi^2\,(d^3-d_1^3)\,(d^2-d_1^2)\text{-}64(d^3\text{-}d_1^3)^2}{1\,152\,\pi\,(d^2\text{-}d_1^2)}$
14. Solid ellipse	$0.785\, bh$	$y = \dfrac{h}{2}$	$\dfrac{\pi\, bh^3}{64}$
15. Hollow ellipse	$0.785\,(bh - b_1 h_1)$	$y = h/2$	$\dfrac{\pi\,(bh^3 - b_1 h_1^3)}{64}$

Section Modulus

$$Z = \dfrac{I}{y} \ \text{or}\ \dfrac{I}{y_1}$$

which one is smaller

Radius of Gyration

$$r_o = \sqrt{\dfrac{I}{A}}$$

Fig. C-1 Forms of Sections

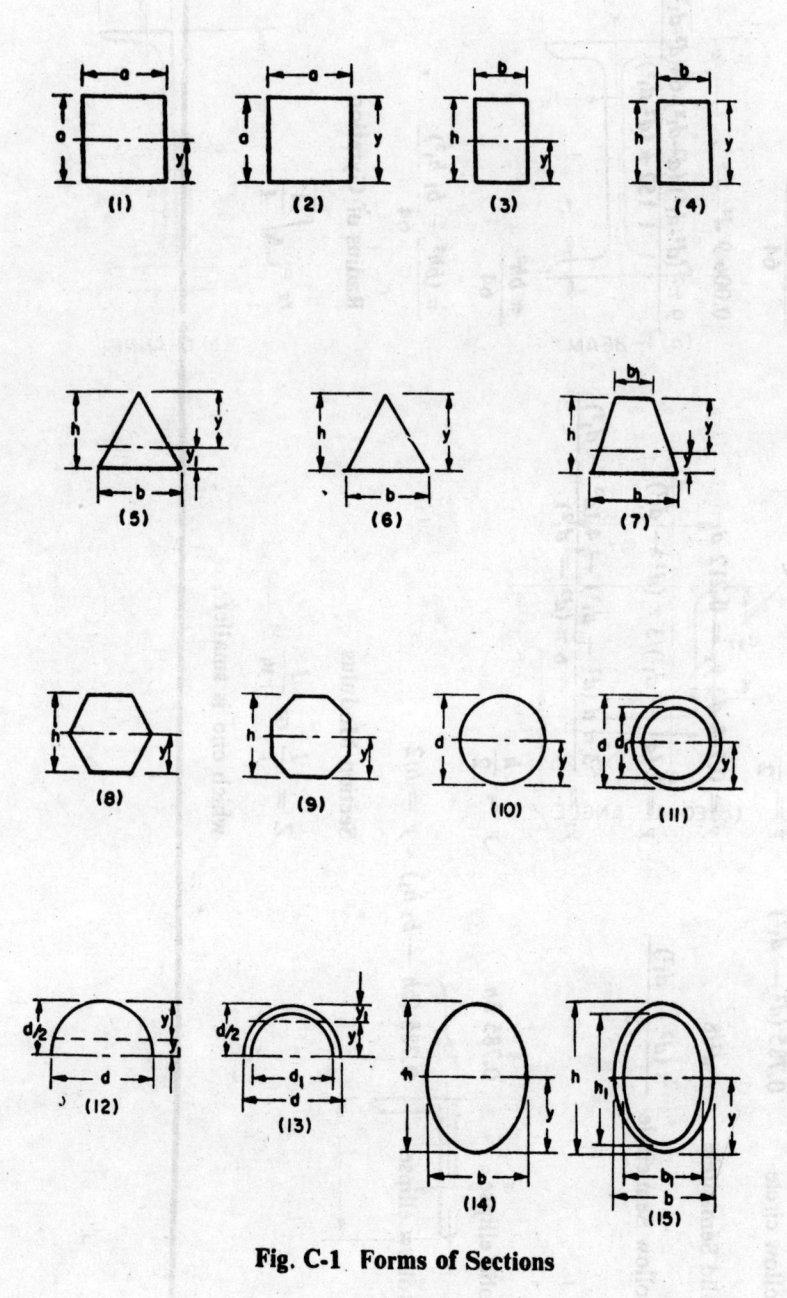

APPENDICES

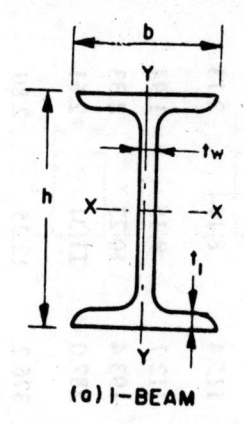

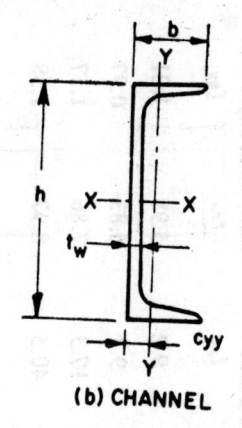

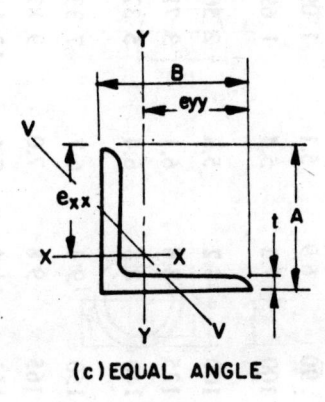

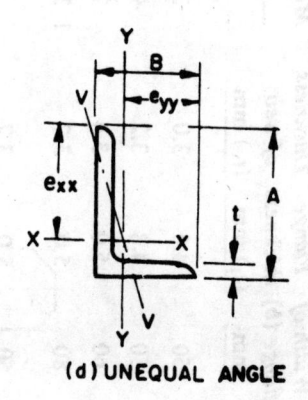

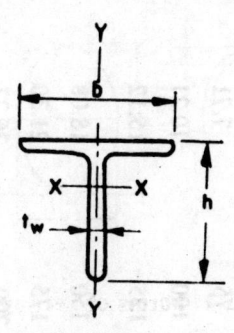

(a) I-BEAM

(b) CHANNEL

(c) EQUAL ANGLE

(d) UNEQUAL ANGLE

(e) TEE BAR

Fig. C-2 Structural Steel Sections

Table C-2 Dimensions and Properties of I-Beams (Fig. C-2a)

Designation		Sectional area (A) cm²	Depth of section (h) mm	Width of flange (b) mm	Flange thickness (t_1) mm	Thickness of web (t_w) mm	Moments of inertia (cm⁴)		Radii of gyration (cm)	
							I_{xx}	I_{yy}	r_{xx}	r_{yy}
ISJB	150	9.01	150	50	4.6	3.0	322.1	9.2	5.98	1.01
	175	10.28	175	50	4.8	3.2	479.3	9.7	6.83	0.97
	200	12.64	200	60	5.0	3.4	780.7	17.3	7.86	1.17
	225	16.28	225	80	5.0	3.7	1 308.5	40.5	8.97	1.58
ISLB	75	7.71	75	50	5.0	3.7	72.7	10.0	3.03	1.14
	100	10.21	100	50	6.4	4.0	168.0	12.7	4.06	1.12
	125	15.12	125	75	6.5	4.4	406.8	43.4	5.19	1.69
ISLB	150	18.08	150	80	6.8	4.8	688.2	55.2	6.17	1.75
	175	21.30	175	90	6.9	5.1	1 096.2	79.6	7.17	1.93
	200	25.27	200	100	7.3	5.4	1 696.6	115.4	8.19	2.13
ISLB	225	29.92	225	100	8.2	5.8	2 501.9	112.7	9.15	1.94
	250	35.53	250	125	8.6	6.1	3 717.8	193.4	10.23	2.33
	275	42.02	275	140	8.8	6.4	5 375.3	287.0	11.31	2.61
ISLB	300	48.08	300	150	9.4	6.7	7 332.9	376.2	12.35	2.80
	325	54.90	325	165	9.8	7.0	9 874.6	510.8	13.41	3.05
	350	63.01	350	165	11.4	7.4	13 158.3	631.9	14.45	3.17

ISLB	400	72.43	400	165	12.5	8.0	19 306.3	716.4	16.33	3.15
	450	83.14	450	170	13.4	8.6	27 536.1	853.0	18.20	3.20
	500	95.50	500	180	14.1	9.2	38 579.9	1 063.9	20.10	3.34
ISMB	100	14.60	100	75	7.2	4.0	257.5	40.8	4.20	1.67
	125	16.60	125	75	7.6	4.4	449.0	43.7	5.20	1.62
	150	19.00	150	80	7.6	4.8	726.4	52.6	6.18	1.66
	175	24.62	175	90	8.6	5.5	1 272.0	85.0	7.19	1.86
	200	32.33	200	100	10.8	5.7	2 235.4	150.0	8.32	2.15
	225	39.72	225	110	11.8	6.5	3 441.8	218.3	9.31	2.34
ISMB	250	47.55	250	125	12.4	6.9	5 131.6	334.5	10.39	2.65
	300	56.26	300	140	12.5	7.5	8 603.6	453.9	12.37	2.84
	350	66.71	350	140	14.2	8.1	13 630.3	537.7	14.29	2.84
	400	78.46	400	140	16.0	8.9	20 458.4	422.1	16.15	2.82
	450	92.27	450	150	17.4	9.4	30 390.8	834.0	18.15	3.01
	500	110.74	500	180	17.2	10.2	45 218.3	1 369.3	20.21	3.52

Table C-3 Dimensions and Properties of Steel Channels (Fig. C-2b)

Designation		Sectional area (A) cm²	Depth of section (h) mm	Width of flange (b) mm	Thickness of web (t_w) mm	Centre of gravity (c_{yy}) cm	Moments of inertia (cm⁴)		Radii of gyration (cm)	
							I_{xx}	I_{yy}	r_{xx}	r_{yy}
ISJC	100	7.41	100	45	3.0	1.40	123.8	14.9	4.09	1.42
	125	10.00	125	50	3.0	1.64	270.0	25.7	5.18	1.60
	150	12.65	150	55	3.6	1.66	471.1	37.9	6.10	1.73
	175	14.24	175	60	3.6	1.75	719.9	50.5	7.11	1.88
	200	17.77	200	70	4.1	1.97	1 161.2	84.2	8.08	2.18
ISLC	75	7.26	75	40	3.7	1.35	66.1	11.5	3.02	1.26
	100	10.02	100	50	4.0	1.62	164.7	24.8	4.06	1.57
	125	13.67	125	65	4.4	2.04	356.8	57.2	5.11	2.05
	150	18.36	150	75	4.8	2.38	697.2	103.2	6.16	2.37
	175	22.40	175	75	5.1	2.40	1 148.4	126.5	7.16	2.38
	200	26.22	200	75	5.5	2.35	1 725.5	146.9	8.11	2.37
	225	30.53	225	90	5.8	2.46	2 547.9	209.5	9.14	2.62
	250	35.65	250	100	6.1	2.70	3 687.9	249.4	10.17	2.89
	300	42.11	300	100	6.7	2.55	6 047.9	346.0	11.98	2.87
	350	49.47	350	100	7.4	2.41	9 312.6	394.6	13.72	2.82
	400	58.25	400	100	8.0	2.36	13 989.5	460.4	15.50	2.81
ISMC	75	8.67	75	40	4.4	1.31	76.0	12.6	2.96	1.21
	100	11.70	100	50	4.7	1.53	186.7	25.9	4.00	1.49
	125	16.19	125	65	5.0	1.94	416.4	59.9	5.07	1.92
	150	20.88	150	75	5.4	2.22	779.4	102.3	6.11	2.21
	175	24.38	175	75	5.7	2.20	1 223.3	121.0	7.08	2.23
	200	28.21	200	75	6.1	2.17	1 819.3	140.4	8.03	2.23
	225	33.01	225	80	6.4	2.30	2 694.6	187.2	9.03	2.38
	250	38.67	250	80	7.1	2.30	3 816.8	219.1	9.94	2.38
	300	45.64	300	90	7.6	2.36	6 362.6	310.8	11.81	2.61
	350	53.66	350	100	8.1	2.44	10 008.0	430.6	13.66	2.83
	400	62.93	400	100	8.6	2.42	15 082.8	504.8	15.48	2.83

Table C-4 Dimensions and Properties of Equal Angles (Fig. C-2c)

Size $A \times B$ mm mm	Thickness t mm	Sectional area (A) cm²	Distance of extreme fibre $e_{xx} = e_{yy}$ cm	Moments of inertia (cm⁴)		Radii of gyration (cm)	
				$I_{xx} = I_{yy}$	I_{vv}	$r_{xx} = r_{yy}$	r_{vv}
20 × 20	3.0	1.12	1.41	0.4	0.2	0.58	0.37
	4.0	1.45	1.37	0.5	0.2	0.58	0.37
25 × 25	3.0	1.41	1.79	0.8	0.3	0.73	0.47
	4.0	1.84	1.75	1.0	0.4	0.73	0.47
	5.0	2.25	1.71	1.2	0.5	0.72	0.47
30 × 30	3.0	1.73	2.17	1.4	0.6	0.89	0.57
	4.0	2.26	2.13	1.8	0.7	0.89	0.57
	5.0	2.77	2.08	2.1	0.9	0.88	0.57
35 × 35	3.0	2.03	2.55	2.3	0.9	1.05	0.67
	4.0	2.66	2.50	2.9	1.2	1.05	0.67
	5.0	3.27	2.46	3.5	1.5	1.04	0.67
	6.0	3.86	2.42	4.1	1.7	1.03	0.67
40 × 40	3.0	2.34	2.92	3.4	1.4	1.21	0.77
	4.0	3.07	2.88	4.5	1.8	1.21	0.77
	5.0	3.78	2.84	5.4	2.2	1.20	0.77
	6.0	4.47	2.80	6.3	2.6	1.19	0.77
45 × 45	3.0	2.64	3.30	5.0	2.0	1.38	0.87
	4.0	3.47	3.25	6.5	2.6	1.37	0.87
	5.0	4.28	3.21	7.9	3.2	1.36	0.87
	6.0	5.07	3.17	9.2	3.8	1.35	0.87

Table C-4 continued

Table C-4 continued

Size $A \times B$ mm mm	Thickness t mm	Sectional area (A) cm^2	Distance of extreme fibre $e_{zz} = e_{vv}$ cm	Moments of inertia (cm^4) $I_{zz} = I_{yy}$	I_{vv}	Radii of gyration (cm) $r_{zz} = r_{yy}$	r_{vv}
50 × 50	3.0	2.95	3.68	6.9	2.8	1.53	0.97
	4.0	3.88	3.63	9.1	3.6	1.53	0.97
	5.0	4.79	3.59	11.0	4.5	1.52	0.97
	6.0	5.68	3.55	12.9	5.3	1.51	0.96
55 × 55	5.0	5.27	3.97	14.7	5.9	1.67	1.06
	6.0	6.26	3.93	17.3	7.0	1.66	1.06
	8.0	8.18	3.85	22.0	9.1	1.64	1.06
	10.0	10.02	3.78	26.3	11.2	1.62	1.06
60 × 60	5.0	5.75	4.35	19.2	7.7	1.82	1.16
	6.0	6.84	4.31	22.6	9.1	1.82	1.15
	8.0	8.96	4.23	29.0	11.9	1.80	1.15
	10.0	11.00	4.15	34.8	14.6	1.78	1.15
65 × 65	5.0	6.25	4.73	24.7	9.9	1.99	1.26
	6.0	7.44	4.69	29.1	11.7	1.98	1.26
	8.0	9.76	4.61	37.4	15.3	1.96	1.25
	10.0	12.00	4.53	45.0	18.8	1.94	1.25
70 × 70	5.0	6.77	5.11	31.1	12.5	2.15	1.36
	6.0	8.06	5.06	36.8	14.8	2.14	1.36
	8.0	10.58	4.98	47.4	19.3	2.12	1.35
	10.0	13.02	4.90	57.2	23.7	2.10	1.35
75 × 75	5.0	7.27	5.48	38.7	15.5	2.31	1.46
	6.0	8.66	5.44	45.7	18.4	2.30	1.46
	8.0	11.38	5.36	59.0	24.0	2.28	1.45
	10.0	14.02	5.28	71.4	29.4	2.26	1.45

80 × 80	6.0	9.29	5.82	56.0	22.5	2.46	1.56
	8.0	12.21	5.73	72.5	29.4	2.44	1.55
	10.0	15.05	5.66	87.7	36.0	2.41	1.55
	12.0	17.81	5.58	101.9	42.4	2.39	1.54
90 × 90	6.0	10.47	6.58	80.1	32.0	2.77	1.75
	8.0	13.79	6.49	104.2	42.0	2.75	1.75
	10.0	17.03	6.41	126.7	51.6	2.73	1.74
	12.0	20.19	6.34	147.9	60.9	2.71	1.74
100 × 100	6.0	11.67	7.33	111.3	44.5	3.09	1.95
	8.0	15.39	7.24	145.1	58.4	3.07	1.95
	10.0	19.03	7.16	177.0	71.8	3.05	1.94
	12.0	22.59	7.08	207.0	84.7	3.03	1.94
110 × 110	8.0	17.02	8.00	195.0	78.2	3.38	2.14
	10.0	21.06	7.92	238.4	96.3	3.36	2.14
	12.0	25.02	7.84	279.6	113.8	3.34	2.13
	15.0	30.81	7.72	337.4	139.3	3.31	2.13
130 × 130	8.0	20.22	9.50	328.3	131.4	4.03	2.55
	10.0	25.06	9.42	402.7	162.1	4.01	2.54
	12.0	29.82	9.34	473.8	191.8	3.99	2.54
	15.0	36.81	9.22	574.6	235.0	3.95	2.53
150 × 150	10.0	29.03	10.94	622.4	249.4	4.63	2.93
	12.0	34.59	10.86	735.4	296.0	4.61	2.93
	15.0	42.78	10.74	896.8	363.8	4.58	2.92
	18.0	50.79	10.62	1 048.9	429.5	4.54	2.91
200 × 200	12.0	46.61	14.64	1 788.9	715.9	6.20	3.92
	15.0	57.80	14.51	2 197.7	883.7	6.17	3.91
	18.0	68.81	14.39	2 588.7	1 046.5	6.13	3.90
	25.0	93.80	14.12	3 436.3	1 411.6	6.05	3.88

Table C-5 Dimensions and Properties of Unequal Angles (Fig. C-2d)

Size A × B mm mm	Thickness t mm	Sectional area (A) cm²	Distance of extreme fibre (cm)		Moments of inertia (cm⁴)			Radii of gyration (cm)		
			e_{xx}	e_{yy}	I_{xx}	I_{yy}	I_{vv}	r_{xx}	r_{yy}	r_{vv}
30 × 20	3.0	1.41	2.02	1.51	1.2	0.4	0.2	0.92	0.54	0.41
	4.0	1.84	1.98	1.47	1.5	0.5	0.3	0.92	0.54	0.41
	5.0	2.25	1.94	1.43	1.9	0.6	0.4	0.91	0.53	0.41
40 × 25	3.0	1.88	2.70	1.93	3.0	0.9	0.5	1.25	0.68	0.52
	4.0	2.46	2.65	1.88	3.8	1.1	0.7	1.25	0.68	0.52
	5.0	3.02	2.61	1.84	4.6	1.4	0.8	1.24	0.67	0.52
	6.0	3.56	2.57	1.81	5.4	1.6	1.0	1.23	0.66	0.52
45 × 30	3.0	2.18	3.08	2.31	4.4	1.5	0.9	1.42	0.84	0.63
	4.0	2.86	3.03	2.27	5.7	2.0	1.1	1.41	0.84	0.63
	5.0	3.52	2.99	2.23	6.9	2.4	1.4	1.40	0.83	0.63
	6.0	4.16	2.95	2.19	8.0	2.8	1.7	1.39	0.82	0.63
50 × 30	3.0	2.34	3.37	2.35	5.9	1.6	1.0	1.59	0.82	0.65
	4.0	3.07	3.33	2.30	7.7	2.1	1.2	1.58	0.82	0.63
	5.0	3.78	3.28	2.26	9.3	2.5	1.5	1.57	0.81	0.63
	6.0	4.47	3.24	2.22	10.9	2.9	1.8	1.56	0.80	0.63
60 × 40	5.0	4.76	4.05	3.04	16.9	6.0	3.4	1.89	1.12	0.85
	6.0	5.65	4.01	3.00	19.9	7.0	4.0	1.88	1.11	0.84
	8.0	7.37	3.93	2.92	25.4	8.0	5.2	1.86	1.10	0.84

APPENDICES

Size									
65 × 45	5.0	5.26	4.43	3.42	22.1	8.6	4.8	2.05	1.28	0.96
	6.0	6.25	4.39	3.38	26.0	10.1	5.7	2.04	1.27	0.95
	8.0	8.17	4.31	3.30	33.2	12.8	7.4	2.02	1.25	0.95
70 × 45	5.0	5.52	4.73	3.46	27.2	8.8	5.1	2.22	1.26	0.96
	6.0	6.56	4.68	3.41	32.0	10.3	6.0	2.21	1.25	0.96
	8.0	8.58	4.60	3.34	41.0	13.1	7.8	2.19	1.24	0.95
	10.0	10.52	4.52	3.26	49.3	15.6	9.5	2.16	1.22	0.95
75 × 50	5.0	6.02	5.11	3.84	34.1	12.2	6.9	2.38	1.42	1.07
	6.0	7.16	5.06	3.80	40.3	14.3	8.2	2.37	1.41	1.07
	8.0	9.38	4.98	3.72	51.8	18.3	10.6	2.35	1.40	1.06
	10.0	11.52	4.90	3.64	62.3	21.8	12.9	2.33	1.38	1.06
80 × 50	5.0	6.27	5.40	3.88	40.6	12.3	7.2	2.55	1.40	1.07
	6.0	7.46	5.36	3.84	48.0	14.4	8.5	2.54	1.39	1.07
	8.0	9.78	5.27	3.76	61.9	18.5	11.0	2.52	1.37	1.06
	10.3	12.02	5.19	3.68	74.7	22.1	13.5	2.49	1.36	1.06
90 × 60	6.0	8.65	6.13	4.61	70.6	25.2	14.3	2.86	1.71	1.28
	8.0	11.37	6.04	4.52	91.5	32.4	18.6	2.84	1.69	1.28
	10.0	14.01	5.96	4.45	110.9	39.1	22.8	2.81	1.67	1.27
	12.0	16.57	5.88	4.37	129.1	45.2	26.8	2.79	1.65	1.27

Table C-6 Dimensions and Properties of Tee Bars (Fig. C-2e)

Designation		Sectional area (A) cm²	Depth of section (h) mm	Width of flange (b) mm	Moments of inertia (cm⁴)		Radii of gyration (cm)	
					I_{xx}	I_{yy}	r_{xx}	r_{yy}
ISNT	20	1.13	20	20	0.4	0.2	0.59	0.39
	30	1.75	30	30	1.4	0.6	0.89	0.57
	40	4.48	40	40	6.3	3.0	1.18	0.82
	50	5.70	50	50	12.7	5.9	1.50	1.02
	60	6.90	60	60	22.5	10.1	1.81	1.21
	80	12.25	80	80	71.2	32.3	2.41	1.62
	100	19.10	100	100	173.8	79.9	3.02	2.05
	150	29.08	150	150	603.8	267.5	4.56	3.03
ISHT	75	19.49	75	150	96.2	230.2	2.22	3.44
	100	25.47	100	200	193.8	497.3	2.76	4.42
	125	34.85	125	250	415.4	1 005.8	3.45	5.37
	150	37.42	150	250	573.7	1 096.8	3.92	5.41
ISST	100	10.37	100	50	99.0	9.6	3.09	0.96
	150	19.96	150	75	450.2	37.0	4.75	1.36
	200	36.22	200	165	1 267.8	358.2	5.92	3.15
	250	47.75	250	180	2 774.4	532.0	7.62	3.34
ISLT	50	5.11	50	50	9.9	6.4	1.39	1.12
	75	9.04	75	80	41.9	27.6	2.15	1.75
	100	16.16	100	100	116.6	75.0	2.69	2.15
ISJT	75	4.50	75	50	24.8	4.6	2.35	1.01
	87.5	5.14	87.5	50	39.0	4.8	2.75	0.97
	100	6.32	100	60	63.5	8.6	3.17	1.17
	112.5	8.14	112.5	80	101.6	20.2	3.53	1.58

REFERENCES :

1. IS : 808—1957 Rolled Steel Beam, Channel and Angle Sections.
2. L. E. Brownell and E. H. Young, "Process Equipment Design", John Wiley and Sons, Inc., New York.
3. R. J. Roark, "Formulas for Stress and Strain" McGraw-Hill Book Company, New York.

INDEX

Advantages of multilayer construction 191
Allowable stresses
    bolting materials 108
    flange 117
Alloys
    aluminium 199
    copper 200
    nickel 200
Aluminium and aluminium alloys 199
Analysis and design
    heads and closures 43
Angles, equal
    dimensions and properties 283, 284
Appendices 261—288
Area, compensation 88
Area, reinforcement 88
Austenitic stainless steels
    mechanical properties 264
Autofrettage 189

Bars
    dimensions and equations 288
    mechanical properties 263, 264
Basic SI units 2
Beam behaviour
    on an elastic foundation 58
Beam, I
    dimensions and properties 280
Beam, semi-infinite 61
Bending moments
    in a bearing plate with gussets 165
    longitudinal, in the vessel shell 177
Bolt area determination 109
Bolting-up condition
    flange moments determination 114
Bolt loads 108, 112
Bolt mateial 107
    allowable stresses 108

Bolt spacing
    recommended 106
    selection 105
Bracket support design 172
Bubble cap tray design 236, 242
    beam support 246
    equations 237
    factors 236
    peripheral tray support ring 245
    sectional trays 240
Butt welded joints details 213

Calculation
    flange 108
    induced flange stress 114
    vessel wall thickness
        design example 191
Carbon and low alloy steels
    mechanical properties 261
Carbon steels 197
Channels, steel
    dimensions and properties 282
Characteristics, materials 197
Clad steels 199
Classification, flanges 100
Classification, pressure 274
Codes, design 13
Coefficient, seismic 149
Coefficient, thermal expansion 268
Coil layer vessel 191
Common supports
    horizontal process vessels 161
    vertical process vessels 161
Compensation
    effective area
        for multiple openings, 96, 97
        protruded nozzle connection 89
        ring pad compensated nozzle opening 90

Compensation
   for multiple openings 96
   for openings
      in head 95
      in process equipment 81—98
         design example 90, 95
      types 82
Conical bottom cylindrical vessel
   design example 254
Conical head 41
   analysis and design 47
   and reducers 41
   stresses 48
   thickness design example 50
Conversion, units 1—11
Copper and copper alloys 200
   mechanical properties 265
   modulus of elasticity 267
Corrosion allowance 18
Corrosion, patterns of 205
Corrosive services
   material specifications 205
Criteria of failure 26
Critical length between stiffeners 129
Cylindrical vessels
   discontinuity stresses 64, 68
   load deformation 61
   reinforcement boundaries 85
   under axially symmetrical loadings 61

Deflection
   bubble cap tray design 241
Derived SI units 2
   with complex names 2
Design
   beam support 246
   bolt load 112
   bracket support 172
   bubble cap tray design 236
   codes 13
   cylindrical and spherical vessels
      under internal pressure 29—37
         design example 36
   equations
      bubble cap tray design 242

   derivation—IS code 31
Design
example
   compensation for opening
      in process equipment 90, 95
   cylindrical and spherical vessels 36
   heads and closures 4, 50, 54, 56
   local stresses in process equipment
      due to discontinuity 66, 69, 72, 75, 77
   nonstandard flanges design 120
   optimization 251, 253, 254, 256
   process vessels and pipes design
      under external pressure 134
   special parts design 229, 242, 244
   support design for process vessels 167
   tall vessels design 153
   thick walled high pressure vessels 191, 193
expansion loop radius 229
flange
   criteria 100
   load 112
   remarks 118
gasket seating stress 103
heads and closures 39—56
   design example 46, 50, 54, 56
loadings 20
lug support 172
multilayer vessel 191
non-standard flanges 99—125
   design example 120
packed bed support plate 246
preliminaries 13—27
process vessels and pipes
   under external pressure 127—139
      design example 134
saddles 182
saddle supports 176
seating stress for gasket 103
sieve plate, perforated 246
skirt bearing plate and anchor bolt 163
skirt support 161
special parts 223—247
   design example 229, 242, 244

INDEX 291

Design
   expansion joint for heat exchangers 223
      design procedure 223
   expansion loop in piping system 226
      design example 229
      expansive forces 227
      induced stresses 228
   stress 16
      factors for various materials 17
   support for process vessels 161—184
      constants for moment calculation 175
      design example 167
   tall vessels 141—160
      design example 153
   temperature 15
   thick walled high pressure vessels 185—195
      design example 191, 193
   tray support ring, peripheral 245
   wall thickness 17
   welded joints 211
Details of butt welded joints 213, 214
Determination
   actual bolt area 109
   area to be compensated 88
   area to be reinforced 88
   bolt area 109
   bold load 108, 109
   compensation requirements for openings in head 95
   cost of optimum storage tank and geometry
      design example 256
   diameter, blank
      dished head 54
   discontinuity stress 66, 69
   equivalent stress under combined loadings 141
   flange moments 113
   fractionating tower shell courses 153
   induced stresses in the pipe 223
   longitudinal stresses 144
   reinforcement boundaries for circular openings in cylindrical and spherical vessels 85
   safe external pressure against plastic deformation 132
   safe pressure against elastic failure 130
   shell thickness of fractionation tower 134
   stress patterns around openings 84
   thermal stresses 75
   thickness
      conical head 50
      dished head 54
      flanged flat cover 46
Diameter
   heat exchanger 272
   plate shells tolerances 129
   process equipment 271
   tube 275
Differential thermal expansion 72
Dilation of pressure vessles 22
Dimensions
   angles, equal 283
   angles, unequal 286
   bars, tee 288
   channels, steel 282
   gasket 105
   I beam 280
   saddles, standard 184
   tank, open rectangular
      design example 253
   tubes, steel 139
Discontinuity
   in process equipment 57
      evaluation 63
   stresses
      design example 66, 69
      due to differential thermal expansion 72
      in cylindrical vessel
         with flat head 68
         with hemispherical head 64
         with vessel wall of dissimilar construction 73
Dished ends opening compensation 96

Dished head
  construction 51
  thickness determination 54
Distillation column
  process design sketch 154
Dye penetrant tests 219

Effective gasket width 104
Effect of concentric load on a beam
  supported by elastic foundation 59
Efficiency factor, weld joint 19
Elastic failure
  safe pressure 130
  theories 186
    thickness of vessel wall
      design example 191
Elastic stability 27
Elevated temperature serivces
  material specification 201
Ellipsoidal opening
  design example 56
Elliptical dished heads 41
  analyses and design 51
Equations, design
  bubble cap tray 237
  expansive forces in pipe lines 227
Equipment fabrication and testing 211–221
Equipment testing 216
Environments
  material specifications 201
Evaluation
  discontinuity stresses in process vessels 63
Expansion bellow 224
Expansion loop
  design example 229
  in piping system 226
External pressure
  pipes and tubes 138
  spherical shell 137

Fabrication and testing, equipment 211–221
Factor, gasket 103

Factor of safety 16
Factors, bubble cap tray design 236
Failure of vessels 127
Ferrous materials
  modulus, of elasticity 266
Fixed tube plate 234
Flange
  calculations 108
  classification 100
  commonly used 99
  integral-type 111
Flange
  load 108
    design 112
    remarks 118
    moment arms 113
  loose type 110
  moments
    determination 113
  narrow faced 100
  nonstandard
    design 99 – 125
    example 120
    criteria 100
  optional type 112
  ring type
    design example 120
  stresses
    subject to external pressure 117
Flanged flat cover thickness
  determination 46
Flanged-only heads 39
Flanged shallow dished heads 40
Flanged standard dished heads 40
Flange stresses 109
Flange subjected to external pressure 117
Flat heads 39
  analysis and design 43
Flexural rigidity — example 75
Forgings
  mechanical properties 261, 264
Formed heads sections 42
Forms of sections 278
Foundation modulus 58

INDEX

Fractionating tower
   bubble cap trays design 242
   plate thickness design example 244
   shell courses design example 153
   shell thickness 134
Freon tests 220
Functions of one variable system
   optimization techniques 250

Gasket
   design seating stress, min 103
   dimensions 105
   materials 103
   selection 102
      temperature and pressure consideration 104
   width
      actual min. 103
      effective 104

Hair-pin U-tube plate 235
Heads
   conical 41, 47
   design 39-56
      example 46, 50, 54, 56
   elliptical dished 41
   flanged only 39
   flanged shallow dished 40
   flanged standard dished 40
      analysis and design 51
   flat 39, 43
   formed
      sections 42
      significance 42
      stress concentration factor 53, 54
   hemispherical 41
   types 39
Heat exchanger, single pass
   shell diameter calculation aid 272
Heat treatment, post weld 216
Hemispherical ends
   opening compensation 96
Hemispherical heads 41
High alloy steels 198
Hydrostatic pressure test 217

I beams
   dimensions and properties 280
Incoloy 201
Inconel 201
Indian Standards on materials 208
Induced stresses in the pipe
   determination 228
Infinitely long beam
   subjected to a concentric load 60
Inspection and nondestructive testing 216
Insulating materials specific weight 153
Integral type flanges 111

Lagrange multiplier 252
Lame's stress analysis
   thick walled cylinder 32
Length
   cylindrical vessels effected by load deformation 62
   tube 275
Limitations
   monoblock high pressure vessel 188
Loads
   bolt 108, 112
   flange 108, 112
Local stresses in process equipment
   due to discontinuity 57—79
      design example 66, 69, 72, 75, 77
Longitudinal stress
   in thin cylinder and sphere 31
Loose type flanges 110
Low alloy steels 198
Low temperature services
   material specification 203
   steels 204
Lug support design 172
Lugs with horizontal plates
   stresses in the vessel wall 173

Magnetic tests 220
Materials
   bolt 107
   Indian Standards 208
   sources of information 210
   specifications 197—210
      for specific environments 201

Maximum principal stress theory 186, 187
Maximum shear theory 186, 187
Maximum strain energy theory 186, 187
Maximum shear theory 186, 187
Mechanical properties
    austenitic stainless steels 264
    bars 263
    carbon and low alloy steels 261
    copper and copper alloys 265
    forgings 261
    high alloy steel 264
    metals 261—269
    pipes 262
    plates 261
    sections 263
    sheet 265
    strip 265
    tubes 262
Metals
    mechanical properties 261—269
    non-ferrous 199
Methods of optimization 249, 250
    analytical 249
    case study 250
    search method 250
Modulus of elasticity
    copper and copper alloys 267
    ferrous materials 266
    nickel and nickel alloys 266
Moment arms for flange loads 113
Moment of inertia 23
Monel 201
Monoblock vessel
    with high strength steels 188
Multilayer construction
    advantage 191
Multilayer vessel design and construction 191
Multiples and sub multiples of SI units 3
Multiplier, Lagrange 252

Nickel and nickel alloys 200
    modulus of elasticity 266
Non-ferrous metals 199

Non-standard flanges
    design 99—125
    example 120
Nozzle opening compensation
    design example 90
Nozzle thickness, minimum 120

Openings
    compensation 96
        effective area 97
        requirement in heads 95
    reinforcing 81
    stress pattern
        theoretical determination 84
    uncompensated 93
Operating condition
    flange moments determination 114
Optimization 249—259
    design example 251, 253, 254, 256
    methods 249, 250
        analytical 249
        case study 250
        search method 250
    techniques 250
        functions of one variable system 250
Optimum diameter to height ratio
    for large oil storage tank
        design example 251
Optional-type flanges 112
Out-of-roundness of shells 129

Period of vibration of the vessel 149
Pipe lines expansive forces 227
Pipes and process vessels
    design under external pressure 127—139
        example 134
Pipes
    mechanical properties 262
    under external pressure
        minimum wall thickness 138
Piping system expansion loop 226
Plates, steel 269
    mechanical properties 261, 263, 264, 265

INDEX

Plate thickness, fractionating column
   design example 244
Pneumatic pressure tests 217
Poisson's ratio 21
   of some metals 269
Post weld heat treatment 216
Pressure
   classification 274
   consideration
      gasket selection 104
   design 14
   external
      pipes and tubes 138
      spherical shell 137
   safe against failure 130, 132
   tests 217
   vessels 22
Pressure vessel, high
   monoblock construction limitations 188
   thickwalled
      design 185—195
      design example 191, 193
Pressure, working 14
Prestressing of thick wall vessels 189
Process equipment
   compensation for openings 81—98
      design example 90, 95
   local stresses
      due to discontinuity 57—79
         design example 66, 69, 72, 75, 77
   nominal diameter 271
Process vessels
   and pipes design
      under external pressure 127—139
         design example 134
   diameter, nominal 271
   support design 161—184
      example 167
   thickness design 36
Properties
   angles, equal 283, 284
   angles, unequal 286
   bars, tee 288

   beams, I 280
   channels, steel 282
   sections 276—288
Proportional limit 128

Radial and hoop stress 30
Radiography tests 218
Radius of gyration 24
Reducers 41
Reinforced vessels under external pressure 128
Reinforcement
   circular openings determination 85
   openings 81
Representation of weldments by symbols 214
Resonant oscillation 149
Ribbon and wire wound vessel 190
Ring stiffeners 181

Saddles, steels
   standard construction 183
   standard dimensions 184
Saddle supports
   design 176
   reinforcing backing plate 181
Screw threads
   system international 270
Search by Golden section
   optimization 255
Sectional tray design 240
Section modulus 24
Sections
   forms 278
   mechanical properties 263, 264
   steel
   structural 279
Seismic coefficient 149
Seismic forces on a tall vessel 148
Selection
   bolt spacing 105
   gasket 102
Semi-ellipsoidal end
   opening compensation 96
Semi-infinite beam 61

Shell, fractionating tower 134
Shell or nozzle flange moment
   stress concentration 78
Shells, out-of-roundness in 129
Shell, spherical
   under external pressure 137
Shell stresses 141
Shell thickness minimum
   petroleum industry 274
   under ordinary conditions 274
   under servere conditions 274
Shrink fitted shell 190
Sieve plate, perforated
   design 246
SI Units
   basic units 2
   conversion factors 8—11
   decimal multiples and submultiples 3
   derived units 2
      with complex names 2
   practical units 6, 7
   prefixes and their symbols 3
   submultiples 3
   symbols 4
Skirt bearing plate and anchor bolt
   design 163
   maximum bending moments 165
Skirt support 161
   design example 167
Skirt wall thickness 162
Sources of information
   material specification 210
Special parts design 223—247
   example 229, 242, 244
Specifications
   material 197—210
      for specific environments 201
      steel 271
      welded joints 211
Specific weight 269
   insulating materials 153
Spherical shell under external pressure 137
Spherical vessels
   reinforcement boundaries 87
Spring constant 58

Standard dimensions 261—288
   flanges 119
Standards, Indian
   materials 208
Standard dished head
   analysis and design 51
Standard values 269—275
   screw threads 270
Steels
   carbon 197, 261
   channels, dimensions, properties 2 2
   clad 199
   high alloy 198, 264
   low alloy 198, 261
   low temperature service 204
   plates 269
   sections, properties 276 -288
   specifications 271
   stainless
      mechanical properties 264
   strip 269
Stiffeners, critical length between 129
Stiffening rings
   design example 134
Storage vessel
   cost—design example 256
   geometry determination 256
   optimum diameter to height ratio
      design example 251
Stress
   allowable
      bolting materials 108
   analysis
      Lame's 32
   axial
      compressive
         due to pressure 144
      tensile
         due to dead loads 144
         due to pressure 144
   bending
      longitudinal 178
         at mid span 178
         at the saddles 178

INDEX

Stress
  circumferential
    design of saddle supports 180
  concentration 24, 84
    factor
      formed heads 53
      theoretical 58
    in the shell or nozzle due to flange moment 78
  discontinuity 63
    cylindrical vessels 64, 68
    due to differential thermal expansion 72
    design example 66, 69
    evaluation in process vessels 63
  equivalent
    under combined loadings 141
  flange 109
  in a thick cylinder 185
  induced
    flange 114
  in the shell 141
  in the vessel wall due to lugs with horizontal plates 173
  local
    in process equipment
      due to discontinuity 57—79
        design example 66, 69, 72, 75, 77
  longitudinal 144, 152
    bending
      due to dynamic loads 145
      due to eccentric loads 152
    in thin cylinder and sphere 31
  patterns around opening 84
  radial — design example 72
  radial and hoop in thin cylinder 30
  seismic loads 148
  tangential 179
  thermal 25
  wind loads 145
Strip
  mechanical properties 265
  steel 269

Structural steel sections 279
Submultiples of SI Units 3
Support design
  bracket 172
  lug 172
  packed bed support plate 246
  process vessels 161—184
    design example 167
  saddle 176
  skirt 161
Symbols 1—11
  welding 214, 215
System international screw threads 270

Tall vessels
  design 141—160
    design example 153
  loads acting 143, 144, 145
  resonant oscillation 149
  under wind pressure 146
Tangential shearing stress
  saddle support 179
Tank, rectangular
  optimum dimension
    design example 253
Techniques of optimization 250
Tee bars
  dimensions and properties 288
Temperature
  design 15
  gasket selection consideration 104
  limitations for pressure parts 16
Tests
  dye penetrant 219
  equipment 211
  freon 220
  magnetic 220
  non-destructive for equipment 216
  pressure 217
  radiography (X-rays) 218
  ultrasonic 221
Theoretical stress concentration factor 58
Theories of elastic failure 186
Thermal expansion coefficient 268

Thermal stresses 25
   determination 75
Thick cylinder stresses 185
Thickness
   dished head 54
   gusset plates 174
   pipe wall
      under external pressure 138
   shell
      ordinary condition 274
      petroleum industry 274
      under severe conditions 274
   skirt-wall 162
   tube plate 231, 234
      minimum 232
      with floating head 235
      with hair-pin U-tube 235
   tube sheet 275
Thick wall vessels, prestressing 189
Thin cylinder
   and sphere
      longitudinal stress 31
      radial and hoop stresses 30
Thin wall vessels 29
Tolerances on diameter of plate shells 129
Torispherical heads
   analysis and design 51
Tube
   diameter and thickness 275
   dimensions
      steel 139
      lengths 275
   mechanical properties 262, 264, 265
   under external pressure 138
Tube plate thickness 231
   fixed 234
   hair-pin U-tube 235
   minimum 232, 275
   with floating head 235
Tube sheet thickness 275
Types
   butt welded joints
      equal thickness 212
      some detail 213
      unequal thickness 214
   compensation for openings 82, 83

flanges, narrow-faced 100
heads 39
   materials 197
      basic characteristics 197
   welded joints 211

Ultrasonic tests 221
Uncompensated openings 93
   design example 95
Unequal angles
   dimensions and properties 286
Units, symbols and conversion factors 1—11

Values
   hub stress correction factor 116
   integral flange factors 115
   loose hub-flange factors 116
   mechanical properties of metals 261
   properties of sections 276
   standard 269
Vessels
   bending moment, longitudinal 177
   common support 161
   failure 127
   multi layer construction
      design example 193
   period of vibration 149
   reinforcement under external
      pressure 128
   tall
      design 141—160
      design example 153

Wall thickness
   design 17
   minimum actual 17
   pipes under external pressure 138
Welded joint
   design 211
   efficiency factor 19
   types and specifications 211
Welding symbols 214, 215
Wind loads in tall vessels 145
Working pressure, minimum 14

X-ray tests 218